SpringerBriefs in Criminology

More information about this series at http://www.springer.com/series/10159

Kandace L. Fisher-McLean

Perceptions of Community Crime in Ferguson, MO

A Qualitative Study Prior to the Death of Michael Brown

Springer

Kandace L. Fisher-McLean
Edwardsville, IL
USA

ISSN 2192-8533 ISSN 2192-8541 (electronic)
SpringerBriefs in Criminology
ISBN 978-3-319-46785-6 ISBN 978-3-319-46786-3 (eBook)
DOI 10.1007/978-3-319-46786-3

Library of Congress Control Number: 2016952890

Printed on acid-free paper

This Springer imprint is published by Springer Nature
The registered company is Springer International Publishing AG
The registered company address is: Gewerbestrasse 11, 6330 Cham, Switzerland

This research is dedicated to the people of Ferguson, Missouri. May your inspiring commitment and enthusiasm illuminate the unfamiliar road ahead as you continue to demonstrate the true meaning of community.

"...it is a neighborhood...There are people that care about one another...A neighborhood is first a place where I live. This is my dwelling place. This is where I rise and shine."

"Loving where you are, loving it with all its quirks. It's hard to explain the feeling...From the very beginning we had felt like we belonged here."

"Community—when you get those first four letters...put them to the side...put a dash after the last "m" you have the word 'unity.' That's what we are. We are unified."

"...we bought this house...we liked the house...When we found out that we loved the community ...we were...beside ourselves...

We were part of...history because this...is an older community."

"I haven't moved more than two yards away in 69 years...I'm either very blessed or very boring...I'm always told that I am very blessed."

Preface

This book is based on a research study conducted in Ferguson, Missouri prior to the shooting of Michael Brown in August 2014. The purpose was to investigate older adult's perceptions of crime in their community and examine how their quality of life was affected. Older adults were chosen specifically for the research study because they previously indicated that they felt crime had increased in their community and no longer felt as safe as they once did. At the time this research study was conducted, reported crime had not dramatically increased. However, during the past 20 years there has been a rapid transformation in many other physical and social aspects throughout Ferguson, possibly interpreted as an increase in crime to longtime community residents. Older adults in Ferguson have a strong connection to their home and community, but as they got older, their community began to change and evolve. They essentially became the only thing remaining constant in their ever-changing world.

This book presents some perspectives about life in Ferguson, Missouri which are much different than what was portrayed through the national media. On the other hand, some resident accounts describe community tensions, racial prejudice, and foreshadowing fears of the events that would occur. In particular, older adult's concerns regarding young people walking down the middle of the street and their apprehension with crime at the Canfield Green Apartment Complex became much more significant after the shooting of Michael Brown transpired.

This research study was not openly accepted throughout the community. Formal complaints were expressed by community members to the researcher and to the City of Ferguson. They feared that this study would draw negative attention to the community. At times, the challenges presented seemed monumental and almost impossible. However, in spite of the opposition, several individuals supported the continuation of this research study; affirming that any student research initiative drawing such community response was worth the perseverance.

My faculty advisor, Dr. Ronald Phillips and committee members guided me through every step of the research process. My friends, family, and colleagues encouraged my work; reading written drafts, discussing the implications of the outcomes, and offering their own insights and perspectives. Finally, this research study would not have been possible without the support of Ferguson Mayor James Knowles III. Before the commencement of this initiative, I did not have a relationship with Mr. Knowles. Given the community attention generated by this research study, he could have asked that I suspend data collection. Instead, he asked to learn more so that he could better educate the residents of Ferguson. He was also interested in learning about resident's perceptions of crime and what the City of Ferguson could do to make their community a safer place to live.

Edwardsville, IL, USA Kandace L. Fisher-McLean

Contents

About the Author

Dr. Kandace L. Fisher-McLean serves as an Associate Professional Housing and Environmental Design Specialist with the University of Missouri Extension based in St. Louis, Missouri. She is also the National Coordinator for the Healthy Homes Partnership. She earned a Ph.D. in Human Environmental Sciences: Architectural Studies from the University of Missouri—Columbia, an M.S. in Human Ecology: Design Studies and a Certificate in Gerontology from the University of Wisconsin—Madison, and a B.S. in Interior Design from Southern Illinois University—Carbondale.

Dr. Fisher-McLean became interested in older adult's perceptions of crime in Ferguson, Missouri in 2009 after receiving a "JumpStart the Conversation" grant to address aging-in-place issues in the community. This project received national attention and was recognized in March 2010 Compendium of Community Aging Initiatives by the Center for the Advanced Study of Aging Services at the University of California at Berkeley—School of Social Welfare. She decided to further investigate older adult's perceptions of crime in Ferguson, Missouri through her doctoral research in 2013 and 2014. To recognize her research in Ferguson, Missouri she was the recipient of the International Architectural Research Centers Consortium King Award for Ph.D. research.

Chapter 1
Introduction

As of 2010, Ferguson, Missouri had a population of 21,203. Ferguson is located in the northern portion of St. Louis County. The total population of St. Louis County is 998,954. The population of St. Louis City adds another 319,294 to the total St. Louis metropolitan area (United States Census Bureau 2010). Figure 1.1 identifies the location of St. Louis County within Missouri to the right and the location of Ferguson within St. Louis County to the left (Arkyan 2007).

Background of the Problem

In July 2009, I was presented with a "JumpStart the Conversation" grant awarded by the Partners for Livable Communities, Metlife, and the National Association of Area Agencies on Aging. As a part of the movement to promote aging-in-place, the St. Louis metropolitan region was selected to host a national workshop. Eight "JumpStart the Conversation" grants were awarded in partnerships with distinct community aging-in-place proposals (Partners for Livable Communities 2014).

My original idea for utilizing the grant money did not fully develop. However, a trusted mentor introduced me to the North County Senior Ministry (NCSM) who was also working to launch an aging-in-place initiative. The thought was that our ideas could be combined to formulate a successful project. NCSM started an initiative that they called the *Lights On! Neighborhood Engagement Program* after identifying two elements of distress for North St. Louis County older adults. These older adults felt that their level of safety was declining and no longer felt connected with their community and neighborhoods. Washington University's Brown School of Social Work partnered with NCSM and verified both fear and decreased engagement among North St. Louis County older adults (Calliotte and Roscano 2009). In their examination of the topic, they discovered that when "individuals perceive a high level of crime and are fearful, they tend to stay inside. However, doing so perpetuates the problem as fewer people are outside to deter actual crime.

K.L. Fisher-McLean, *Perceptions of Community Crime in Ferguson, MO*,
SpringerBriefs in Criminology, DOI 10.1007/978-3-319-46786-3_1

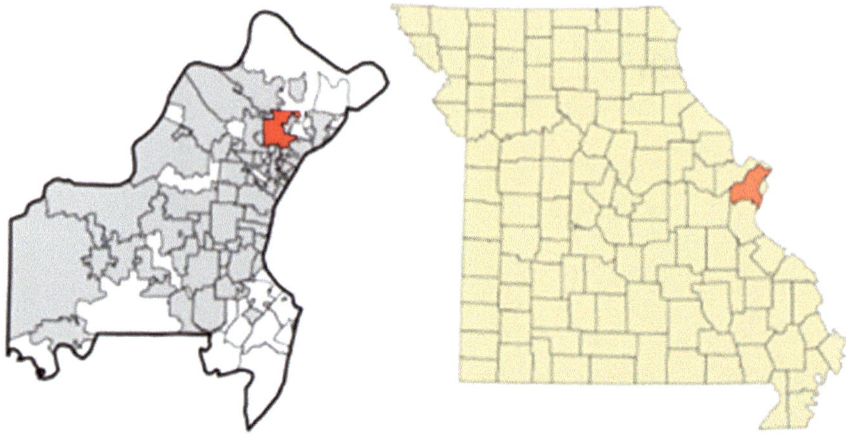

Fig. 1.1 Map illustrating location of Ferguson and St. Louis County in Missouri. Retrieved from Wikipedia, 2014. Created by Arkyan (2007)

Their research also showed that older adults report the highest level of fear of crime" (Calliotte and Roscano 2009).

After a review of several municipalities in North St. Louis County, Ferguson, MO was chosen for the initiative (Communications, NCSM, August 2009). Local law enforcement determined that motion sensor lighting and landscape pruning could be offered to older adults. Keeping shrubbery neatly trimmed, below window height increases visibility for homeowners and neighbors and decreases areas where criminals can hide around homes (Calliotte and Roscano 2009). A representative from NCSM made a presentation to the Ferguson Neighborhood Improvement Program (FNIP) board. As a result, they committed to sponsoring 20 homes with motion sensor lights with the caveat that the installation occur in an area of Ferguson with higher reported crime. After consulting with the then current Ferguson Neighborhood Watch Coordinator, the Southwest Neighborhood in Ferguson was identified to launch the initiative (Communications, NCSM, August, 2009).

In addition to motion sensor lights, it was decided that additional funds from the "JumpStart the Conversation" grant could be utilized for solar powered address signs, dead bolt locks with new lever handles for those residents who needed it, and skid proof/reflective tape for difficult concrete steps at the entrances to homes. Local police officers stated that they had trouble seeing resident's address numbers in the event of an emergency and the addition of lighted address signs would help.

The initiative lasted approximately 1 year and involved 18 older adults in the Southwest neighborhood of Ferguson, Missouri. With the help of a social work practicum student working with NCSM, a survey was conducted and identified that one-half of residents in the initiative had either been a victim of crime or someone in their household had been a victim of crime. Three-quarters of the residents knew of a neighbor who had been a victim of crime. At the completion of the program, an ice cream social and trivia night were hosted in an effort to promote social engagement among residents.

This program received national attention, in part because of being a "JumpStart the Conversation" recipient. It was recognized in the March 2010 Compendium of Community Aging Initiatives by the Center for the Advanced Study of Aging Services at the University of California at Berkeley—School of Social Welfare.

As we entered homes and spoke with residents regarding their safety concerns, I became curious about the "bigger" community issue. Older adults repeatedly stated that they felt crime was increasing in their neighborhood and that they were more fearful than they had ever been. However, at the same time, they were expressing their desire to want to age-in-place in their current home, neighborhood, and community. Many of the older adults involved in the initiative were longtime residents and seemed to have a deep connection with their home and surroundings.

Following the conclusion of the initiative, I thought repeatedly about what could be done to further understand resident's concerns and improve their quality of life. Digging deeper, I found that reported crime had not increased significantly in Ferguson during the past 20 years (Federal Bureau of Investigation United States Department of Justice 2013; Dr. Bob Bursik, personal communication, 2011). However, what had evolved during the past 20 years were changes in various physical and social factors such as the community demographics and increases in rental, foreclosed, and abandoned and neglected homes. Therefore, it seemed that the rapid progression of these changes within the community could symbolize crime to longtime residents. They fear that the changes taking place and interpret their community is generally going downhill. Given the changes, older adults in Ferguson might also perceive that their quality of life has diminished if they socially withdrawal from activities because of safety fears.

Contextual Data

Below are a series of charts outlining reported crime, changes in racial demographics, and rental versus owner occupied housing in Ferguson, Missouri. This supports the notion that crime has not increased dramatically, while other physical and social factors have changed (Tables 1.1, 1.2 and 1.3).

Purpose of the Study

Understanding how older adult's perceptions of crime in Ferguson affected their quality of life was the focus of this research inquiry. Specifically, older adults were asked to explore, with the use of personal photography, some of the physical and social factors influencing their perception of crime in the community. The intention

Table 1.1 Number of reported violent crimes and number of reported property crimes between 1990 and 2010 in Ferguson, Missouri

Year	Months	Population	Violent crime total	Property crime total
1990	12	22,286	69	662
1991	12	22,464	95	994
1992	12	22,616	112	1018
1993	12	22,216	122	1175
1994	12	22,402	104	1360
1995	12	22,783	107	1133
1996	12	22,932	116	1118
1997	12	23,115	97	1113
1998	9	21,264	86	978
1999	12	20,602	91	1337
2000	12	22,406	102	1341
2001	2	22,544	12	285
2002	12	22,716	93	1466
2003	12	22,256	108	1399
2004	12	22,100	87	1268
2005	12	21,849	108	1109
2006	12	21,615	134	1062
2007	12	21,142	105	1208
2008	12	20,967	163	1291
2009	12	20,814	113	1171
2010	12	21,203	94	1121

United States Department of Justice: Federal Bureau of Investigations (2013)

Table 1.2 Demonstration of changes in racial demographics between 1990 and 2010

Years	Population	White number	White percentage (%)	Black number	Black percentage (%)	Other	Other percentage
1990	22,286	16,454	73.80	5589	25	243	1
2000	22,406	10,026	44.70	11,743	52.40	637	2.8
2010	21,203	6206	29.3	14,297	67.4	700	3.3

United States Census Bureau (1990–2010)

Table 1.3 Demonstration of changes in occupied, vacant, owner occupied, and renter occupied property between 2000 and 2010

Years	Occupied	Vacant	Owner occupied	Renter occupied
2000	8612 (93.7 %)	579 (6.3 %)	5745 (66.7 %)	2867 (33.3 %)
2010	8192 (90 %)	913 (10 %)	4891 (59.7 %)	3301 (40.3 %)

United States Census Bureau (2000 and 2010)

was to better understand and eventually help reduce fear of crime, improve the rate of those wishing to age-in-place, and improve the lives of older adults in Ferguson and other communities.

Significance of the Study

Data collection for this study concluded in early June 2014, nearly 2 months before the fatal shooting of Michael Brown by a police officer in Ferguson, Missouri. The heartbreaking incident sparked community outrage, protests, violence, prayers for justice and peace, and distrust between a predominately White police force and predominately Black residential population in Ferguson.

Unfortunately, most of the nation, through the lens of the media, witnessed only a snippet of life in Ferguson, suspended in time. This research presents some perceptions of Ferguson that are much different than what was portrayed in national headlines after the death of Michael Brown in August 2014. On the other hand, some residents describe details that foreshadow the devastating events occurred. Despite the tragedy, Ferguson is still a small town where people work, play, shop, pray, grow-up, and grow old. It is home for the people who dwell there and have made Ferguson their residence of choice. Perceptions are very powerful shaped by the phenomena one experiences through every fiber of their beings. Perceptions are also highly susceptible to change. No argument will be made that Ferguson is anything less than a passionate community; as witnessed through this study and the events that unfolded during and after August 2014.

Theoretical Framework

Numerous theories developed a framework to understand older adult's perceptions of crime. First, Lawton's Environmental Docility Theory (1974), states "the less competent the individual in terms of personal disability or deprived status, the more susceptible is his behavior to the influence of immediate environmental situations." This explains why older adults are more likely to be affected by environmental stressors such as crime, especially if they experience diminished physical and cognitive skills and feel they are unable to ward off an attacker.

The Broken Window Theory explains how minor community damage, if left unrepaired, can escalate into serious decay and eventually significant social deterioration. As Kelling and Wilson (1982) describe, "if a window in a building is broken and left unrepaired, all the remaining windows will soon be broken." One unrepaired window sends a message to those passing by that no one cares, invites others to break more windows, and can ultimately lead to serious social breakdowns and community decay (Kelling and Wilson 1982).

The Social Disorganization Theory offers some of the "most promising research on fear of crime" and "stresses the relationships among demographic characteristics, social integration, and fear of crime" (Adams and Serpe 2000). According to Kornhauser (1978, p. 120) and Bursik (1984, p. 12), "social disorganization is the failure of community structure to appreciate the shared values of its residents and sustain effective social controls" (cited by Sampson and Groves 1989). "The hypothesis of the Social Disorganization Theory is that low economic status, ethnic heterogeneity, residential mobility, and family disruption lead to community social disorganization which in turn increases crime and delinquency rates" (Sampson and Groves 1989).

Oscar Newman's Defensible Space Theory explains how environmental factors can hinder crime, "by creating the physical expression of a social fabric that defends itself" with "real and symbolic barriers, strongly defined areas of influence, and improved opportunities for surveillance—that combine to bring the environment under the control of the residents" (Newman 1972, p. 3). A product of Defensible Space Theory, Newman (1972) describes, crime prevention through environmental design (CPTED) as the "design or redesign of an environment to improve safety, to decrease incidents of criminal activity, and eliminate conditions that may contribute to crime" (cited by Walsh 1999). Strategies include "territoriality (sense of ownership), surveillance, access control, and activities promoted within spaces including higher densities and mixed use development in order to optimize the number of potential eyes on the street" (Cozens 2007).

Additionally, Relph's (1976) Theory of Place Attachment describes, "place attachment as the authentic and emotional bond with an environment that satisfies a fundamental human need" (cited by Scannell and Gifford 2010). As cited by Morgan (2010), "…where a person lives in a particular locale over an extended period, that person will often develop feelings of affection for, and a sense of belonging, or being of that place, so that place becomes 'one anchor of his or her identity' (Hay 1998)." Although some older adults in Ferguson have expressed that they do not feel as safe as they once did, they feel an attachment to their home, neighborhood, and community. This is where they grew up, bought their first home, raised their children, and are retiring.

Methodological Framework

A qualitative research approach in which purposive sampling, autophotography, photoelicited interviews, and a short survey were used in data collection for this research study. Transcendental phenomenology was utilized to understand and interpret older adults' perceptions of crime in Ferguson, Missouri. Through this process, the researcher sets aside all prejudgments and prior knowledge of the subject, known as epoche. Next, the researcher analyzes accounts from research participants to identify important statements. These statements are paired down and grouped into themes. The themes are used to develop descriptions of the phenomenon and are further utilized to assemble the final essence of the experience (Moustakas 1994; Moerer-Urdahl and Creswell 2004).

Qualitative methods were utilized as opposed to quantitative methods in this research to gain a deeper understanding of older adults' perceptions of crime in their community. It is hoped the research results can be used to develop an action plan to alleviate some concerns that older adults have and improve their quality of life. The study results will not be generalizable to older adults in other communities. However, it is felt that the methods developed can be employed in other communities to gain a deeper understanding of particular community issues.

Results Synopsis

Fifteen older adults from 8 of the 12 neighborhoods within Ferguson, Missouri participated in the research study between October 2013 and June 2014. Based on the photographs and the interviews, 12 themes emerged: community image, interaction with nature, undesirable attitudes and behaviors, littering, historic buildings and homes, community safety, community investment, vacant property, racial integration and community transformation, natural surveillance, maintenance, and rental property. Photographs taken by the research participants were paired with their quotes to formulate a complete representation of each thematic area.

Synthesis of the Experience

Maintenance or "to maintain" becomes the over-arching experience extracted from the composition of themes. Older adults who participated in the research study desired to maintain some control over the physical and social image of their community as a means for preserving the perceived shared values of the community. While the physical and social image of the community rapidly transformed, older adults struggled to make sense of the fact that their control and tolerance for these changes has slipped away.

Discussion

Older adults who participated in this research were less concerned with personal victimization and more concerned with various signs of physical and social deteriorations that could indicate more serious long-term community decay. However, at least for these older adults, their extraordinary investment and social integration in the community seemed to counter some of the fears they could experience from crime and community deterioration. In describing disorder in their community, older adults begin to foreshadow some of the events leading up to the shooting of Michael Brown in August, 2014; making strong connections with the Attorney General Report released subsequent to the shooting and community riots.

References

Adams, R. E., & Serpe, R. T. (2000). Social integration, fear of crime, and life satisfaction. *Sociological Perspectives, 43*(4), 605–629.

Arkyan. (2007). *Map illustrating location of Ferguson and St. Louis County in Missouri.* Retrieved from http://en.wikipedia.org/wiki/Ferguson,_Missouri

Calliotte, W., & Roscano, E. (2009). *Lights on!*. Ferguson Times: Collaborative brings safety to neighborhoods in North County.

Kelling, G. L., & Wilson, J. Q. (1982, March). Broken windows: The police and neighborhood safety. *Atlantic Monthly*, 29–38.

Lawton, M. P. (1974). Social ecology and the health of older people. *The American Journal of Public Health, 64*(3), 257–260.

Moerer-Urdahl, T., & Creswell, J. (2004). Using transcendental phenomenology to explore the "ripple effect" in a leadership mentoring program. *International Journal of Qualitative Methods, 3*(2). Retrieved from http://www.ualberta.ca/~iiqm/backissues/3_2/pdf/moerer.pdf

Morgan, P. (2010). Towards a developmental theory of place attachment. *Journal of Environmental Psychology, 30*, 11–22.

Moustakas, C. (1994). *Phenomenological research methods.* Thousand Oaks, CA: SAGE Publications Inc.

Newman, O. (1972). *Defensible space: Crime prevention through urban design.* New York, NY: Macmillan.

Partners for Livable Communities. (2014). *St. Louis jumpstart grantees.* Retrieved from http://livable.org/program-areas/livable-communities-for-all-ages-a-aging-in-place/the-aging-in-place-initiative/jumpstart-the-conversation-grants

Relph, E. (1976). *Place and placelessness.* London: Pion.

Sampson, R. J., & Groves, W. B. (1989). Community structure and crime: Testing social disorganization theory. *American Journal of Sociology, 94*(4), 774–802.

Scannell, L., & Gifford, R. (2010). Defining place attachment: A tripartite organizing framework. *Journal of Environmental Psychology, 30*(1), 1–10.

United States Census Bureau. (1990–2010). *American fact finder.* Retrieved from http://factfinder.census.gov/faces/tableservices/jsf/pages/productview.xhtml?src=CF, http://factfinder.census.gov/faces/tableservices/jsf/pages/productview.xhtml?src=CF, and http://censtats.census.gov/cgi-bin/pl94/pl94data.pl

United States Census Bureau (2000 and 2010). *American fact finder.* Retrieved from http://factfinder.census.gov/faces/tableservices/jsf/pages/productview.xhtml?src=CF

United States Department of Justice: Federal Bureau of Investigations. (2013). *Uniform crime reporting statistics.* Retrieved from http://www.ucrdatatool.gov/Search/Crime/Local/JurisbyJuris.cfm?CFID=33479888&CFTOKEN=60320537

Walsh, E. (1999). Crime prevention through environmental design. *Journal of Housing and Community Development, 56*(4).

Chapter 2
Literature Review

The progression of transportation, the construction of the interstate system, and suburban development have all had an incredible influence on the evolution of Ferguson and surrounding suburbs. As these suburbs have aged, they have faced significant demographic shifts, an aging and inexpensive housing stock, as well as various other physical, social, and economic changes. The dramatic subprime lending crisis of the mid-2000s has and will continue to have a significant impact on the area as well.

Ferguson: Suburban Development

Transportation

Railroads and streetcars established growth to the suburbs of St. Louis, Missouri and greatly influenced Ferguson's origins (Fox 1995, p. 186); producing a hotbed for freight and passenger railroad transportation at the close of the nineteenth century (City of Ferguson 2014; Fox 1995, p. 186). In 1876, a spur line connected Ferguson to St. Louis' downtown Union Depot, transforming its rural feel into a suburb attractive for wealthy estates (Fox 1995, p. 186; City of Ferguson 2014). Streetcar suburbs expanded prior to 1918, making a ten-mile commute from downtown accessible in 30 min and "laid the skeleton for the new emerging metropolitan area…" (Ames 1995).

The development of the earliest automobile suburbs occurred between the end of World War I and the end of World War II (Ames 1995). At the close of World War II, those residing in Ferguson preferred to utilize personal automobiles rather than trains and streetcars (Fox 1995). Through the Federal Aid Highway Act, 41,000 miles of limited access interstates connected major cities across the country

K.L. Fisher-McLean, *Perceptions of Community Crime in Ferguson, MO*,
SpringerBriefs in Criminology, DOI 10.1007/978-3-319-46786-3_2

between 1956 and 1972. Interstate 70 was the first project in the St. Louis area to be completed, connecting the downtown area to the airport and Northwest St. Louis County into St. Charles County (Missouri Department of Transportation [MODOT] 2013). Ferguson is geographically bound by Interstate 70 on its south, Interstate 170 on its west, and Interstate 270 on its north.

The automobile became one of the most expansive influences on "large-scale suburbanization" (Ames 1995) and the Federal Aid Highway Act has affected "every aspect of American life in the past 50 years" (MODOT 2013). Suburbs were places where people could "escape the congestion and clamor of the city" (Fox 1995, p. 161), but their development via construction of the interstate highway system drove downtowns into decline, destroyed historical buildings, and cut paths through "vibrant, working class neighborhoods" (MODOT 2013). Post-World War II became known as the era of the "freeway suburbs" (Ames 1995) due to the construction of the "interstate highway system, the rise of the automobile, and the introduction of government home-buying incentives in the 1940s and 1950s" (Fox 1995, p. 161).

Real Estate Expansion

St. Louis City remained a major industrial player throughout World War II. However, by the 1950s there were issues with blight, congested traffic, crime, and businesses and families moving into suburbs throughout St. Louis County (Hannon 1986). Active real estate development turned farmland into sprawling suburbs after WWII and such towns as Florissant, Missouri, a neighboring suburb of Ferguson, saw a spike in their local population. In 1950, the population of Florissant was 3737. By the mid-1970s, 18,000 homes were built and the population peaked at 65,908 (Hannon 1986; Fox 1995, p. 36). Ferguson also boomed after World War II and homes were built by the masses. During the 1940s and 1950s, simple two bedroom homes were constructed throughout Northeast St. Louis County (Duncan 2008). New arrivals to the area worked at McDonnell Aircraft, Universal Match, and Emerson Electric (Fox 1995). Figure 2.1 exhibits a house typically constructed during this era that can be found throughout Ferguson and surrounding suburbs.

Veteran's returning home from World War II were in need of housing and qualified for "guaranteed mortgages under the Veteran's Administration Housing Program" (Fox 1995, p. 36). This program required only a small down payment and thus made home ownership more affordable than renting (Fox 1995, p. 36). After World War II, St. Louis City "had no more room to grow" (The City of St. Louis, Missouri 2011) and "thousands of families left the aging row houses and apartments of their city neighborhoods for lawns and ranch houses" in the suburbs (Fox 1995, p. 36).

Fig. 2.1 Example of brick house constructed Post-WWII

Inner Ring Suburbs

Ferguson, Missouri is considered an inner ring suburb because it is one of the older suburbs close to the City of St. Louis. Inner ring suburbs are also called first suburbs because they were established in the years following World War II (Duncan 2008). They have endured significant demographic and economic changes (Duncan 2008) and "have evolved into places with varied characteristics, assets, and problems" (Hanlon 2009). In the *Death and Life of Great American Cities*, Jacobs (1961) explains that today's worst cities were once dignified suburbs. "…today's brand-new suburbs or semisuburbs are going to be engulfed in cities and will succeed or fail in that condition depending on whether they can adapt to functioning successfully as city districts" (p. 16). According to Lewyn (2000), "suburban sprawl, like the French Revolution, devours its own children. Sprawl creates inner ring suburbs, only to destroy them a few decades later by creating outer suburbs to skim off their elites. So as long as cities and older suburbs continue to lose their most affluent citizens to newer suburbs, no community is truly safe from the ravages of neighborhood decay, and no stable community can endure." Some inner suburban sections are more attractive than others but "there are very few cities that are homogenous in terms of race, socio-economic status, or class" (Fox 1995, p. 161). In fact, Hanlon (2009) suggests, "it has been demonstrated that, in general, declining suburbs are more likely to be inner ring rather than outer suburbs (Leigh and Lee 2005; Lee 2005; Lucy and Phillips 2000, 2006; Hanlon 2008)."

"Mobility and migration have splintered many communities that were long based on blood ties, cultural heritage and life within a narrow geographical region" (Roseland 2005, p. 154). As a result of this mobility, migration, and suburban development, "metropolitan areas countrywide are seeing similar patterns of development-explosive sprawl where farmland and open space once reigned, matched by decline and abandonment in the central cities and older suburbs" (Dionne 1998, p. 66). Ferguson is far from being a brand-new St. Louis suburb as newer suburbs stretch further west to the far edges of St. Louis County into the Missouri River Valley. West is also where the population of St. Louis County appears to be headed. Inhabitants of the area utilize the interstate highway system to zip past older, less desirable suburbs and quickly to houses in subdivisions that are new and sprawling with flashy attractions, malls, and restaurants. Ferguson's housing stock is older and less expensive and as a result younger newer minority families are moving in and taking advantage of the more economical, smaller, living arrangements. It also seems that landlords also take advantage of this inexpensive, smaller housing stock by converting homes into rental property.

Racial Segregation

The automobile and the interstate highway system were not the only shapers of suburban development throughout the St. Louis area. Large-scale racial segregation also contributed greatly to the suburban evolution during the middle of the twentieth century, creating future irreversible housing conditions. "The geography of America would be unrecognizable today without the racist social engineering of the mid-20th century" (Coates 2014). In the 1950s, many St. Louis African–Americans historically lived in decaying slums until the population of the city was fragmented by a fleeing White population and the collapse of industrial development (Gay 2014). The Pruitt–Igoe housing development is one of the most infamous urban renewal projects in history (Gay 2014). Pruitt was designed for Blacks and Igoe was designed for Whites. However, both Pruitt and Igoe filled with Blacks because Whites were more interested in less expensive options available in the suburbs (Rothstein 2014b). Pruitt Igoe was ultimately a catastrophic failure inundated by crime and maintenance issues; three of the high-rise buildings were imploded in 1972 (Hoffman 2000). Pruitt Igoe was located approximately 11 miles from Ferguson, Missouri.

The decay of the city quickened as urban renewal projects paired with massive highways which carved patterns through long established black areas of the community. The innermost suburbs such as Ferguson absorbed these residents as they were displaced (Gay 2014). This resulted in White residents fleeing to more affluent outer-ring and other predominately White suburban areas. Even though it is unconstitutional to discriminate based on race, St. Louis has been plagued with "geographic, cultural, and economic" segregation (Cooperman 2014). Citing the 2010 United States Census, Cooperman (2014) suggests that "North of Delmar Boulevard, St. Louis neighborhoods are about 95 % Black. South of Delmar, they are almost two-thirds White (and the median household income is $25,000 higher). White flight and Black flight adhere to this pattern: Whites ended up in South (St. Louis) County and Blacks in North (St. Louis) County."

According to Rothstein (2014a), suburban development throughout the St. Louis area as well as the nation banned African–Americans through the utilization of community covenants mutually agreed upon by White homeowners and with "racially neutral zoning rules" designating outer-ring suburbs for the wealthy. This left inner-ring suburbs, such as Ferguson to Black homeowners who were escaping the failing schools in the inner city. As Massey and Denton explain in their book, ***American Apartheid*: *Segregation and the Making of the Underclass***, "some real-estate agents reaped huge profits by blockbusting: scaring Whites out of neighborhoods with the threat of Black neighbors, then selling their vacant homes to African-Americans at rapacious prices with predatory loans" (cited by Kovarik 2008). Rothstein (2014b), also explains that policies were purposefully designed by the government to isolate racial minorities. This was accomplished through zoning that produced regional slums, the design of nonintegrated public housing units, White only residential improvements with racially motivated neighborhood restrictions, and an overall deprivation of essential amenities to Black neighborhoods which further solidified their status as a slum to White communities.

By the 1960s, virtually the entire home real estate market was impacted by federal civil rights laws and discriminatory and segregated housing practices did not seize. The Federal Housing Administration (FHA) began a program in 1968 targeting low-income homeowners and "signaled the federal government's larger commitment to subsidizing low income families and individuals rather than the projects in which they might live (Gordon 2008)." Gordon (2008, p. 118) acknowledges that theoretically, these types of initiatives were capable of distributing homeowners receiving housing assistance throughout the geographic area, but in actuality "they simply laundered federal subsidies through the deeply discriminatory institutions of private realty" in which "money and helpless buyers were simply fed into the segregated market" (USCCR 1971; Orfield 1981). Subsequently, Gordon (2008, p. 119) affirms that Black buyers were shown homes in North St. Louis and inner-ring Northern suburbs while Whites were shown homes in predominantly White suburban areas. Housing and Urban Development (HUD) interrupted FHA sales in the transitional Skinker-deBaliviere neighborhood after resident were concerned "that federal subsidies were tipping the local housing market by making mortgages available 'to some gal out of Pruitt Igoe.'" (Leven 1976; USCCR 1971; Orfield 1981). Many of these actions concluded in the late 20th century but "continue to determine today's racial segregation patterns" (Rothstein 2014a) and will continue to have irreversible everlasting effects on housing, suburb development, and the history of St. Louis and our nation.

Ferguson, like much of the rest of North St. Louis County has experienced a shift in the demographics of their community. For example, the population of those 65 and older in Ferguson and other areas surrounding the Lambert International Airport area decreased between 1990 and 2000 (OSEDA 2013). According to OSEDA (2013), younger African American populations have replaced the older Caucasian populations in the area. In Ferguson, the racial composition of the community has changed significantly during the past 20 years. In 1990, Whites made up 73.8 % of the population and Blacks made up 25.1 % of the population. In 2010, Whites made up 29.3 % of the population and Blacks made up 67.4 % of the population (United States Census Bureau 1990–2010). In the 1970s, 97 % of the population in the Northeast Suburbs were White. By the 1990s, middle income Black homeowners migrated to these communities as the first generation of older

White adults moved out (Duncan 2008). Although Ferguson lies just outside the Northeast suburban boundary, it possesses a similar housing stock, has experienced the effects of foreclosure, and a change in the demographics of their community.

Picture Pleasant Suburbia

"Ferguson is a picture pleasant suburbia, a town of tree lined streets and well-kept homes, much of them built for the middle class at mid-century" (Gallagher 2013). Unfortunately, it also took a major hit during the subprime lending crisis of the mid-2000s. Between 2004 and 2007, more than half of the mortgages were subprime in the inner ring suburbs, creating a "cycle of financial destruction for residents and neighborhoods" (Gallagher 2013). Property values declined and the tax base was diminished by foreclosures (Lloyd 2012). Home prices throughout St. Louis have begun to stabilize and some have begun to rise. However, in North St. Louis County prices remain weak and there is an abundance of foreclosed homes that are older, smaller, and out of fashion; selling to landlords who then convert them into rental homes (Gallagher 2013). "Metro St. Louis is a national hot spot for 'underwater' mortgages…and the problem is particularly acute in north St. Louis County" (Gallagher 2014). According to Gallagher (2014), those who are underwater are "roughly twice as likely as others to default on their mortgages, leading to foreclosure. Some argue that stressed homeowners are less likely to improve their homes, or even maintain them, and that can affect the surrounding neighborhood." Some, such as William Rogers, at the University of Missouri-St. Louis argue that, "Homeowners who can't leave have a special interest in keeping the neighborhood nice" and "they might become better neighbors with more civic involvement" (cited by Gallagher 2014).

Ferguson was also struck by a tornado on Good Friday in 2011. According to the National Weather Service Weather Forecasting Office (NWSWFO 2012), the tornado was an EF 4; the strongest tornado St. Louis County has seen in 44 years. This tornado destroyed 251 homes in St. Louis County and thousands more were damaged (Currier 2011). Almost exactly 2-years later, in 2013, another round of tornadoes struck the St. Louis area causing damage in Ferguson, Missouri (NWSWFO 2013). The combined destruction of these tornadoes is still seen in damaged and abandoned homes.

Although Ferguson has experienced significant changes throughout its life cycle, the pride citizens have for their city and the responsibility that they feel for others in their community and future generations is intensely evident. As some would suggest, North St. Louis County seems to be consistently targeted by negative media publicity as it relates to crime, education, and the state of housing throughout the area. Unfortunately, St. Louis in general has received national attention about its crime rate. The City of St. Louis has consistently received high rankings as one of the most dangerous cities in America. However, one of the reasons for this is that unlike other growing cities, St. Louis has not grown beyond its restricted borders.

Other cities have expanded their borders, pulling in safer surrounding neighborhoods, thus diluting the overall crime statistics for the area. If St. Louis were to do this, the crime rate would be much lower (Christie 2013).

Older Adults and Fear of Crime

Minnery (1986) explains that fear of crime is problematic when it restricts the activities of individuals so gravely that it disturbs their quality of life (Garofalo 1977; Lavrakas 1982; Maxfield 1984) and can greatly affect such groups of people as the elderly (e.g., Goldsmith and Goldsmith 1976; Markson and Hess 1980). Psychologist Werner Greve presented empirical data that indicated a similar degree of fear of crime in youth and elderly, even though it is perceived that the elderly exhibit more fear of crime. However, young and old people seem to manage victimization risk differently. Older adults' inclination is to perceive danger and be accordingly careful. Youth weigh the cost versus benefit of a potential dangerous situation (Herz and Kania 2002).

Taylor and Covington (1993) found that residents felt that their daytime safety was compromised when they perceived that issues could ascent from unsupervised groups of youth in their community. In past conversations, one older resident in Ferguson stated that she walked in a local park regularly. However, an incident arose in which she was approached by a group of African American teenagers. Even though there was no confrontation and no harm was done, this sufficiently scared the older woman to a point where she no longer returned to the park for her regular walks.

Chadee and Ditton (2003) reexamined a study conducted by Ferraro and LeGrange (1992) confirming in their findings that there is no relationship between age and fear of crime. However, Markson and Hess (1980) cite a study conducted by Harris et al. (1975) that fear of crime varies with age and gender. "At every age, women have a greater fear of crime than do men; 20 % of all women compared to 11 % of all men. This fear increases with age-from 10 % among women 18–54 years old, 18 % among those 55–64 years old, and 28 % among those 65+." The research varies greatly on whether older adults have a higher fear of crime in comparison to other age groups. However, because older adults initially expressed that they felt crime was increasing, it became important to understand how older adults perceive crime in their community and the potential effect fear has on their quality of life.

Perceived Environmental Press and Individual Competence

It appears that crime has not statistically increased in Ferguson, however, as previously noted, there have been significant physical and social changes occurring in the community. It could be argued, therefore, that some of these changes symbolize a perceived increase in crime by older adults even if actual crime is not increasing. Older adults who continue to reside in the community appear to have a sufficiently high enough competency level to deal with the psychological demands placed upon them from fear of crime. This higher level of competency potentially translates into a higher tolerance level developed from the gradual physical and social changes taking place in their community. In the past, they have been able to sufficiently "adapt to" or "tolerate" the physical and social changes happening in their community. However, as they age and perceive that their biological and mental competency level is diminishing, their ability to continually "adapt to" or tolerate the evolving physical and social changes in their community may also diminish.

As presented by Lawton (1974), the Environmental Docility Theory states that "the less competent the individual in terms of personal disability or deprived status, the more susceptible is his behavior to the influence of immediate environmental situations." The Environmental Docility Theory is demonstrated through the Environmental Press Model seen in Fig. 2.2. As interpreted by Pollack and Patterson (1980), "through the adult life, a state of balance must exist between the individual's competence…and environmental press…for that individual to maintain a satisfactory quality of life." Due to the reduction of biological and psychological

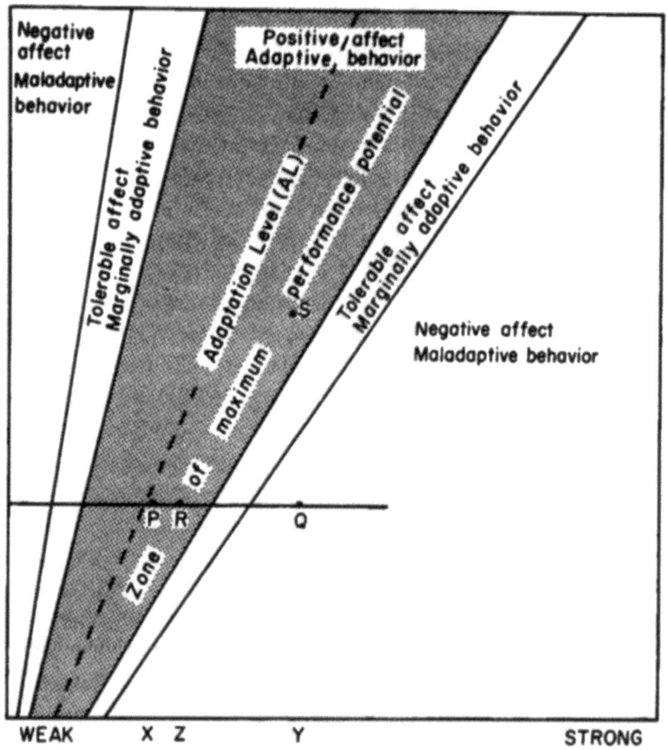

Fig. 2.2 Lawton and Nahemow (1973)

capabilities, one's competency decreases and unless environmental demands are reduced, they will be unable to continuously adapt to their surroundings. If older adults in Ferguson perceive the physical and social changes (environmental demand) taking place as an increase in crime, their ability to adjust may be exceeded. This enhances the perception that they will be victimized and increases the chance that they will begin to withdraw from the community.

Nahemow and Lawton (1973) suggest that "an individual is operating as his best when the environmental press are moderately challenging. If the environment offers too little challenge, the individual adapts by becoming lethargic and thus operates below his capacity." This is a common occurrence in situations where an older adult is moved to a long-term care facility where staff performs all essential tasks for them. "On the other hand, it may be that the environment is too stressful and he has adapted by turning off…When the environmental press are very strong the individual may panic and attempt to escape from the field either physically or psychologically. This escape-oriented behavior is not geared to dealing with the situation represented by the immediate environment, but it is adaptive in the sense that it may remove the person from an intolerable situation" (Nahemow and Lawton 1973).

Because of the perceived increase in crime that older adults in Ferguson suggested was making them more fearful, it is possible that some have attempted to physically escape by moving away from the community. Those who are unable to physically relocate have attempted to cope with their situation by socially and psychologically escaping. Although there are a number of factors that could be associated with an older adult leaving the community, fear of crime could pose an environmental stress contributing to a feeling that their only way of successfully adapting is to escape to somewhere where they feel safer. For older adults who have relocated, the feeling of safety may imply a community with greater demographic homogeneity, economic stability, and fewer signs of physical decay such as litter, uncared for lawns, and run down homes. For those who have chosen to stay in Ferguson despite their perception of crime, they may have retreated socially and psychologically as a method of removing themselves from a situation they can no longer tolerate. Adams and Serpe (2000) uncover "that as social disorder increases in a neighborhood, people report less helping behavior among neighbors and less satisfaction with their neighborhood. In other words, crime is related to social isolation, which in turn contributes to fear of crime and perceived vulnerability (Skogan 1990; Lewis and Salem 1986; Thompson and Norris 1992)."

As cited by Pollack and Patterson (1980),

> Neugarten and Associates (1964) suggest that there is "a shift in personality to less 'active mastery' in later years. As people age, they view themselves as being less capable of dealing with problems through their own initiative. Similarly, Lawton et al. (1976) have discussed a lack of environmental control as partially accounting for the crime problem among the elderly. They note such problems among the elderly as (a) limits in their visual and auditory acuity which result in potential threats going unrecognized; (b) physical limits to the actions they can take such as running to avoid assault; and (c) social isolation and economic deprivation which lead to vulnerability."

All of these factors can contribute to older adults' perceptions that crime has a greater impact on them than in reality, thus making them feel less capable in their individual competency. According to Hale (1996, p. 95), "people who feel unable to protect themselves, either because they cannot run fast, or lack the physical prowess to ward off attackers, or because they cannot afford to protect their homes or because it would take them longer than average to recover from material or physical injuries might be expected to fear crime more than others" (cited by Cossman and Rader 2011).

Social Connectedness and Social Isolation

According to Adams and Serpe (2000), "social integration refers to a person's sense of belonging or attachment to the community (Adams 1992a; Kasarda and Janowitz 1974)" and activities that socially connect people to their community contribute to feelings that they have a common bond with those around them, feeling that they are an important asset to their community, and overall increases their satisfaction with the neighborhood. Hunter and Baumer (1982) note that "social integration into the local community reduces the number of strangers in the neighborhood, reinforces the feeling that neighbors will offer assistance when needed, and contributes to perceptions that the neighborhood is a safe place to live" (cited by Adams and Serpe 2000).

In contrast, when individuals become socially isolated they become more fearful, feel more vulnerable, and withdrawal from the community (Adams and Serpe 2000). Atkins et al. (1991) acknowledge that there is an acute issue associated with fear of crime and suggest that it affects "travel and activity patterns, constrains participation in social activities, generates psychological stress, and is arguably a severe limitation on individual liberty." Miethe (1995) suggests that "fear is functional to a society when it reduces individuals' exposure to risky and vulnerable situations, but fear becomes totally dysfunctional when it permanently restricts and limits all other aspects of social life."

Additionally, as cited by Bellair (2006) "when a large proportion of community residents live in fear, behavioral patterns that emerge from daily interaction among community residents, such as surveillance of space, may be stunted (Goodstein and Shotland 1980)." Informal surveillance occurs when residents are out walking around their neighborhood, interacting with their neighbors, and showing general concern for one another. However, fear of crime may greatly restrict one's participation in neighborhood and social organizations that initiate informal surveillance, such as neighborhood watch groups. Criminals then prey on factors of distress, low internal locus of control, feelings of vulnerability, activity restriction, and a disconnection of social relationships to carry out criminal acts (Bellair 2006).

From earlier discussions with older adult residents in Ferguson, it appeared that those who were more socially integrated into neighborhood watch groups, church activities, and other volunteer activities felt that they had access to more resources when they were faced with a particular problem. As a result of attending the neighborhood watch meetings, residents reported that they felt they had a closer, more intimate relationship with local police officers and did not hesitate to give them a personal call when they felt it was necessary. In contrast, one resident who no longer drove a car and rarely got out of her home stated that she did not know her neighbors well and would not even know who to go to for help if she needed it. Having access to a diverse array of connections and resources becomes increasingly important as one becomes older and begins to rely more heavily on others to assist in day-to-day needs.

Social Support Networks

"Social factors, specifically social support, are important buffers to the stressors of later life…" (Blazer 2005). Aging without the social support network to ease perceived environmental demands from crime may potentially become too difficult to adapt to as their window of competency diminishes. Thompson and Krause (1998) argue that "elders who are embedded in active social networks tend to have better physical and mental health than older adults who are less involved with other people." Thompson and Krause (1998) describe social support as a multidimensional construct developed by Barrera (1986) and is "measured through social embeddedness (the frequency of contact with others), received support (the amount of tangible help that others actually provide), and perceived support (subjective evaluations of supportive exchange, such as satisfaction with support and expectations of future support). Studies increasingly suggest that perceived support may have a stronger effect on health and well-being than other types of social support measures. This is particularly true of anticipated support, which is defined as the belief that others will provide assistance in the future should the need arise." Anticipated support could come from past support that was received or through the involvement in a variety of social networks that the older adult feels could assist them in times of crisis (Thompson and Krause 1998).

Among older adults in Ferguson, the mere thought that someone would be there if necessary could be sufficient in helping combat the fears they experience. This is potentially why the local neighborhood watch meetings attract so many older adults. This venue provides a caring environment in which the local police officer in charge pays special individualized attention to their needs and concerns. He also provides them with a variety of resources and knowledge that they would otherwise be unaware of. Neighborhood watch meetings become a social gathering for the older adults where they have opportunities to share stories with one another, combating the potential psychological fear they feel. Yin (1980) observes that the presence of a vast social support network helps to reduce fear in the elderly because the catastrophic element of victimization is diminished if the elderly feel they can more easily recover with the help of those around them. Merry (1976) also suggests that "those who knew the people they encountered in their neighborhood and were familiar with its norms have a lower degree of fear of crime. Those who regarded their neighbors as 'strangers' were more fearful" (cited by Yin 1980). Yin (1980) refers to a study conducted by Cohn et al. (1978) that found that "those who were involved in a community crime prevention program were less fearful than those who were not…First, joining a community crime prevention program allows access to the social structure of one's neighborhood. Neighbors, therefore, cease to be strangers. Second, participation in any community program may foster development of one's social support network; thus perceived ability to recuperate from a victimization experience is increased." It is hoped that active engagement in community organizations, church groups, and neighborhood watch groups will socially engage residents and help reduce their fear of crime in neighborhood settings. As cited by Adams and Serpe (2000) and Schieman (2009), people who perceive that they have a connection with those in their neighborhood and regularly interact with one another see their neighborhood as more secure (Hartnagel 1979; Lewis and Salem 1986; Riger et al. 1981).

In contrast, Yin (1980) suggests that "interaction with friends and neighbors about crime (regardless of personal experience) constitutes another probable source of fear." Clarke (1983) reports that neighborhood watch groups are largely unsuccessful as a method of community surveillance because people do not sustain interest and are unlikely to express vigilance to protect others. This presents the contradiction associated with neighborhood watch meetings. It seems that several older adults in Ferguson attend the neighborhood watch meetings because they believe it is helpful in understanding crime in their community. However, in addition to the social support provided through the neighborhood watch meetings, they are also made highly aware of every crime related event in their community. The police officer who leads the meetings provides those in attendance with a detailed list of all crime happenings in Ferguson during the past month. The crimes are listed by type, date of occurrence, time of occurrence, and the street where the crime occurred. Having access to this additional information about crime in the community may serve as a significant explanatory factor in influencing older adults' increased perception of crime in their community.

Ross and Jang (2000) found that "living in a neighborhood with a lot of perceived disorder significantly affects mistrust and fear of victimization…Perceived neighborhood disorder and social ties significantly interact: informal social ties with neighbors reduce the fear—and mistrust—producing effects of disorder. However, formal participation in neighborhood watch organizations shows little buffering effect." Informal social ties with neighbors leads to informal surveillance of the individual's neighborhood because residents have a reason to walk around the neighborhood, interact with one another, and show general concern for one another. Through the process of informal surveillance, residents are much more likely to identify disorder in the neighborhood.

As coined by Jacobs (1961) in her book, *The Death and Life of Great American Cities*, having "eyes on the street" is important to keep social order in check. "Public peace-the sidewalk and street peace-of cities is not kept primarily by the police, necessary as police are. It is kept primarily by an intricate, almost unconscious, network of voluntary controls and standards among the people themselves, and enforced by the people themselves" (pp. 31–32). As fear takes over, citizens physically and socially withdrawal into their homes and away from the community and consequently, informal surveillance decreases. Signs of physical disorder advance as criminals create mayhem, assuming citizens are apathetic about what happens in their community, and no one cares enough to observe their delinquent actions (Sampson and Raudenbush 2004).

The Social Disorganization Theory

The Social Disorganization Theory explains the relationship between social changes and perceptions, and how it relates to one's thoughts about crime in their community. According to Adams and Serpe (2000) and Schieman (2009) "much of the most promising research on fear of crime is based on social disorganization theory, which stresses the relationships among demographic characteristics, social integration, and fear of crime. In particular, the theory suggests that people who feel attached to their neighbors also perceive the neighborhood as safer" (Hartnagel 1979; Lewis and Salem 1986; Riger et al. 1981). Social engagement occurs in many forms; being involved in various community organizations, church groups, volunteer opportunities, employment opportunities, or through socialization with friends and neighbors in the community. The presence of social networks also helps provide formal and informal surveillance over neighborhood issues and helps older adults feel more deeply rooted in their community, increasing their overall life satisfaction.

Sampson and Groves (1989) suggest that the hypothesis of the social disorganization theory is that "low economic status, ethnic heterogeneity, residential mobility, and family disruption lead to community social disorganization, which, in turn, increases crime and delinquency rates." Kornhauser (1978, p. 120) and Bursik (1984, p. 12) state, social disorganization is the failure of the community to maintain "effective social controls" and "realize the common values of its residents" (cited by Sampson and Groves 1989). Sampson and Groves (1989) also report, "the structural dimensions of community social disorganization can be measured in terms of the prevalence and interdependence of social networks in a community—both informal (e.g., friendships ties) and formal (e.g., organizational participation)—and in the span of collective supervision that the community directs toward local problems (Thomas and Znaniecki 1920; Shaw and McKay 1942; Kornhauser 1978)."

Ethnic Heterogeneity and Crime Perception

In considering the many social variables that may be associated with older adults' fear of crime, the Social Disorganization Theory explores how ethnic heterogeneity is a contributing factor in crime and fear of crime in the community. Skogan (1995) presented results from surveys finding that "residential proximity to Black people is related to Whites' fear of crime."

This is also evident in the history of Ferguson, Missouri. Kinloch was a predominately Black suburb neighboring a nearly all White Ferguson. The primary thoroughfare between Ferguson and Kinloch was chained off to bar most African–Americans from entering Ferguson up until the mid-1960s (Rothstein 2014b). As Ferguson became more integrated with African Americans, affluent Whites began to flee to other outer ring suburbs. As cited by Quillian and Pager (2000), "Taub et al. (1984) found that individuals respond most strongly to perceived neighborhood crime and housing deterioration in determining when to move and where to settle" and that "Whites may avoid neighborhoods with many Black residents not because of an aversion to neighbors who are Black but because Black neighborhoods on average have higher rates of neighborhood problems like higher crime rates and dilapidated housing stock (Frey 1979; Taylor 1981; Liska and Bellair 1995; Liska et al. 1998)."

Despite Whites appearing to be more afraid, Blacks are overall more fearful of crime due to a concentration of other factors such as differences in victimization, social disorder, and neighborhood physical decay (Skogan 1995). Older Black adults may have a good reason to be more fearful. According to statistics reported by the United States Department of Justice (2012), elderly Blacks "experienced violent victimization at five times the rate of elderly Whites or Asian and Pacific Islanders." In a study conducted by Wiltz (1982), Black older adults were fearful of crime on a daily basis and "those who were victimized had a higher fear of crime compared to those who were not victims" (cited by Joseph 1997).

Socioeconomic Status and Crime Perception

As part of the pilot study conducted in Southwest Ferguson during the winter of 2012, 100 % of older adult men and 71 % of older adult women surveyed strongly agreed or agreed they felt that an increase in unemployed people in their community was related to crime. Although participants were not asked to expand upon their feelings, it seems possible they felt that those who were unemployed might be more desperate than those with employment and a steady income, and therefore more likely to be prone to criminal actions. Americans often think about crime and poverty synonymously. People become more fearful when they feel that their safety is threatened by those of a lower economic status. Even worse than the challenge to their personal safety, people become apprehensive when they deem that their moral, cultural, and economic principles are being eroded (Gans 2013).

Although those of a perceived lower socioeconomic status may be targeted as crime generators, they themselves may also be more fearful of becoming a victim of crime. In Austin et al.'s (2002) study about the effects of neighborhood conditions and perceived safety, they refer to work done by Austin et al. (1994) in which they identified a "significant positive relationship between education and feeling of perceived safety." They also point to studies conducted by Lee (1981), Skogan and Maxfield (1981), and Toseland (1982) in which they found that "higher status was associated with lower levels of fear." Additionally, as cited by Bennett et al. (2007), urban lower income residents from groups of racial and ethnic minorities are most likely to rate their neighborhood as unsafe (Boslaugh et al. 2004; Wilson et al. 2004; Brownson et al. 2001). However, as Austin et al. (2002) indicate, the thought that those of a lower socioeconomic status are more fearful of crime is not collective. They refer to a study performed by Baba and Austin (1989) which found that socioeconomic status had "no significant impact on perceived levels of neighborhood safety."

The Broken Window Theory

Areas with multiple physical incivilities such as litter, graffiti, broken windows, unkempt lawns, and homes in disrepair suffer consequences beyond those of the immediate physical decay and disorder. These physical characteristics become important symbols that residents and others cannot or will not protect their neighborhood from crime (Kelling and Wilson 1982; Skogan 1990). Kelling and Wilson (1982) identify what is known as the Broken Window Theory: "if a window in a building is broken and left unrepaired, all of the remaining windows will soon be broken." One unrepaired window signals to those passing by that no one cares about the building and invites others to break more windows. This can apply to other various situations, such as abandoned cars, empty lots littered with trash, and graffiti. If one person dumps trash in the empty lot next door and no one complains about it or cleans it up, then it must not concern people. This might invite others who have had similar ideas to also use the empty lot as their personal dumping grounds. As referred to in Herzog and Flynn-Smith (2001), how people care for their environment also connects with the Broken Window Theory and "draws support from the wide range of studies implicating maintenance as a predictor of fear of crime (e.g., Perkins et al. 1992, 1993; Schroeder and Anderson 1984; Taylor et al. 1985; Wilson 1975; Wilson and Kelling 1982)." Brown et al. (2004), suggest "residents react to the symbolism of these incivilities by withdrawing from social activity in the neighborhood." According to Taylor et al. (1985), "signs of neighborhood decay make residents infer that their community is going downhill and that nobody is doing or can do anything about it and residents begin to fear for their own safety." From a criminology perspective, this can lead to serious social breakdowns in the community. "A piece of property is abandoned, weeds grow up, and a window is smashed. Adults stop scolding rowdy children; the children, emboldened, become more rowdy. Families move out and unattached adults move in" (Kelling and Wilson 1982). This explains the process simply, but essentially describes how physical and social disorder can quickly snowball, resulting in residents migrating out of the community.

The Defensible Space Theory

Newman's (1972) book, *Defensible Space: Crime Prevention Through Urban Design* identifies the key strategies of "territoriality (sense of ownership), surveillance, access control, and activities promoted within spaces including higher densities and mixed use development in order to optimize the number of potential eyes on the street" (cited by Cozens 2007). Newman's defensible space principles are useful in addressing the needs of the older adults in Ferguson because they address not only the built environment, but its interaction with sociological explanations for crime (Jacobs 1961; Newman 1972). Defensible space is produced "by creating the physical expression of a social fabric that defends itself" with "real and symbolic barriers, strongly defined areas of influence, and improved opportunities for surveillance—that combine to bring the environment under the control of the residents" (Newman 1972, p. 3).

Defensible space principles alone are insufficient in addressing the negative perceptions that older adults have about crime in their community, but it is one of a combination of approaches that can be taken to help alleviate the fears they experience. One major theme of the Defensible Space Theory is to empower residents to feel as if they have control over their physical environment. By utilizing defensible space principles to take control of the built environment, it is also hoped that residents will feel socially and psychologically empowered; easing the fears they experience.

Territorial Reinforcement

Newman (1972) describes territoriality as the "capacity of the physical environment to create perceived zones of territorial influences" (p. 51). According to the Lancaster Community Safety Coalition (2013), territorial reinforcement is "the use of physical attributes that express ownership such as fences, signage, landscaping, lighting, pavement design, etc." Signs of territorial reinforcement can be direct or subtle. A more direct form would include putting a fence around a yard. A subtler form might include landscaping to help define a space and express territory or changes in paving material from the public sidewalk to the private walking path of one's yard. According to Newman (1972), "by its very nature, the single family home is its own statement of territorial claim. It has defined ownership by the very act of its positioning on an integral piece of land buffered from neighbors and public street by intervening grounds" (p. 51). Newman (1972) states that territory can be "reinforced by symbolic shrubs or fences, and in other cultures by high walls and gates" (p. 51), but as housing becomes denser, the territorial definition becomes difficult (p. 52).

A significant issue in Ferguson is that older adults have felt like there was nothing they could do to alleviate crime related problems. Residents can be provided with knowledge and skills to protect themselves from criminal activity. Some methods of territorial reinforcement may be feasible for older adults to apply to their home and surrounding property. Other methods may be more difficult or less feasible to apply. Carefully designed landscaping around homes in Ferguson may be a simple approach to define territory and provide a symbolic transitional edge between public and private space. Some streets in residential neighborhoods within Ferguson have no sidewalks between public streets and private residential yards. Constructing sidewalks to define territory, to provide paths for natural access control, and to provide opportunities for natural surveillance is a sizeable project that would need to be systematically organized and carried out by the City of Ferguson.

However, what should be avoided, and what Newman did not intend to happen is for territorial reinforcement to become a form of "target hardening." Target hardening refers to the use of locks, bars on windows, and gates as an obtrusive version of territorial reinforcement. Although these measures may be effective in keeping potential offenders out, they can negatively alter the mental psyche of someone who is already fearful of crime in their neighborhood. As cited by Wortley (1996), "concerns have been raised about the social implications of the unfettered application of target-hardening principles" (Bottoms 1990; Grabosky 1994; Weiss 1987). Wortley (1996) argues that such features as "walls, guards, conspicuous security devices and the like divide rather than build communities by separating and isolating their members" and "run the danger of becoming counterproductive, creating the very social conditions which foster criminal behavior." Territorial reinforcement principles should rely on subtle measures that convey a similar message, but that also blend in more aesthetically with the built and natural environment. Measures of territorial reinforcement should not be utilized to provoke more fear among residents thus creating additional social isolation for those living in the community.

Natural Surveillance

Newman (1972) defines natural surveillance as the "capacity of physical design to provide surveillance opportunities for residents and their agents" (p. 78). The Lancaster Community Safety Coalition (2013) further explains this concept by describing natural surveillance as the "placement of physical features, activities, and people in a way that maximizes visibility. A potential criminal is less likely to attempt a crime if he or she is at risk of being observed. At the same time, we are likely to feel safer when we can see and be seen." Environmental design attributes that provide natural surveillance include accessible sidewalks, sufficient residential scale lighting, and landscaping appropriately trimmed so that it is easier to detect suspicious activity. In addition, homes and businesses should have direct views to the primary street through doors and windows for easy observation (Lancaster Community Safety Coalition 2013). It is important to note, that depending on the literature type, different phrases are used to describe surveillance. In the Criminology literature the phrases, "informal" and "formal" surveillance are used to describe the ways in which people keep an eye on the neighborhood. Formal methods are carried out by police, security guards, and surveillance cameras. Informal methods can be carried out by wives and husbands, older adults, children, and pedestrians. In the Environment and Behavior literature, surveillance is often referred to in the form of natural surveillance carried out by residents as part of their everyday life; giving less attention to formal methods of surveillance.

As explained by Reynald and Elffers (2009), opportunities for crime depend on the types of activities that occur in a place, the environmental characteristics of place, the motivation of an offender, and the absence of surveillance (Sherman et al. 1989). Newman (1972) explains that "certain sections and arteries of a city have come to be recognized as being safe—by the nature of the activities located there; by the quality of formal patrolling; by the number of users and extent of their felt responsibility; and by the responsibility assumed by employees of bordering institutions and establishments. The areas most usually identified as safe are heavily trafficked public streets and arteries combining both intense vehicular and pedestrian movement; commercial retailing areas during shopping hours, institutional areas; and government offices" (pp. 108–109).

As cited by Sampson and Groves (1989), social organization or disorganization is dependent on informal (friendships) and formal (organizational involvement) social groups in the community and work to supervise local problems in the community such as crime and delinquency (Thomas and Znaniecki 1920; Shaw and McKay 1942; Kornhauser 1978). As some older adults are retired and potentially spend more time at home and in the immediate neighborhood, they become a vital form of natural surveillance because they have the ability to keep a watchful eye on the neighborhood when others may not be home to do so. Recognizing and empowering older adults to act as an important form of neighborhood surveillance provides them with an additional purpose to stay actively invested in the interest of their neighbor and community.

Environmental Perception of Place

The perceptions that people have of their own environment can greatly affect the way that they feel about crime in their neighborhood and community. "The environment…is a language of communication, with a syntactic and semantic structure" (Appleyard 1979). In other words, the environment, while highly structured, is also full of potential meaning, symbols, and interpretations (Appleyard 1979). The way that individuals interpret the structure of their neighborhood is so powerful that it can dictate the entire way they experience it. As people experience their neighborhood, they may interpret what they observe in the form of social symbols and meaning. The social meanings attached to the environment become extremely critical when a perceived conflict occurs in the community or environment. One's environment, especially one's home and immediate neighborhood, are symbol-laden. The home environment, over which one has more direct immediate control, evolves as one evolves, taking on one's personality as one changes, adapts, and grows. However, outside the door, the immediate neighborhood is uncertain territory because individuals may feel they have little control over their neighbors, their neighbors' actions, and what generally happens in the streets of their community (Appleyard 1979). Fear may ensue if people begin to perceive their neighborhood is deteriorating in a physical and/or social sense and there is little they feel they can do to control the circumstances.

For the older adults in Ferguson, they and their homes have become the constant, when the outside world seems to be transforming physically and socially. Their interpretation of the deteriorating physical condition (litter, un-maintained homes, and unkempt lawns) of their neighborhood and the social changes (change in racial composition, population age, and socioeconomics) taking place become potential symbols of community decline and crime. As previously noted, as people's ability to adapt decreases, their ability to continually adjust to the physical and social changes taking place in their community diminishes. Unless the environmental press placed upon them can be reduced, a satisfactory quality of life will likely not be able to be maintained (Lawton 1974; Pollack and Patterson 1980). As older adults begin to experience increased environmental press from changes in the physical and social condition of their neighborhood and their window of adaption decreases, they may believe that there is little that they can do to improve the environment outside of their home. As they retreat into the familiarity of their home and isolation increases, social relationships suffer, and natural surveillance throughout the neighborhood decreases.

Fear of Crime and Threats to Placemaking

During previous conversations with older adults in Ferguson, one question seemed particularly relevant: "why do these people want to stay here if they are so frightened by crime that they do not want to walk in the local park by themselves?" However, this is because older adults' situations were viewed from an outsider's perspective because their sense of place was not understood. Relph (1976) describes, "place attachment as the authentic and emotional bond with an environment that satisfies a fundamental human need" (cited by Scannell and Gifford 2010). As cited by Morgan (2010) described place attachment as "the experience of a long-term affective bond to a particular geographic area and the meaning attributed to that bond. Where a person lives in a particular locale over an extended period, that person will often develop feelings of affection for, and a sense of belonging, or being of that place, so that place becomes 'one anchor of his or her identity' (Hay 1998)." Although some of the residents no longer felt as safe in their community as they once did, their community still felt like the comforts of home after 30–50 years of residence there. One resident displayed all of the tick marks on her door where she had tracked the height of her children over the years. This was the moment when their connection to home and community despite their fears became apparent.

Relph (1976) would describe this connection as the degree of "insideness" that someone has or feels for their community. Insideness is the intensity that people feel in a particular place that forms their identity with that place. It refers to the "degree of attachment, involvement, and concern that a person or group has for a particular place." People will have a stronger identity with place if they feel that they are more inside a place. People can also feel separated or alienated from a place for a given reason, this experience of a place is known as "outsideness." If one feels outside a place, they perceive an emotional separation between them self and that place. Relph (1976) explains that there are degrees of intensities of the insideness and outsideness that people feel for place because different places take on different feelings and meanings based on their experiences and interactions in a place (cited by Seamons and Sowers 2008).

Older adults in Ferguson may have once felt that they were very "inside a place" when roaming through their community. However, this degree of "insideness" is threatened if one fears for his or her personal safety because they perceive that crime is increasing in the community. They may begin to feel more like emotional "outsiders" as their community begins to rapidly change around them and their personal world begins to shrink. To continue as "insiders" in the community older adults must maintain their social integration, personal relationships, and attachments within the larger community.

As part of identifying with a place, Cuba and Hummon (1993) recognize that "community attachment research indicates that integration into the local area is a prime determinant of attachment to locale. Local social involvements-particularly those with friends, but also those involving kin, organizational memberships, and local shopping-prove to be the most consistent and significant sources of sentimental ties to local places (Gerson et al. 1977; Goudy 1982; Guest and Lee 1983; Hunter 1974; Kasarda and Janowitz 1974; St. John et al. 1986)." Social engagement for the older adults in Ferguson is especially critical to staying connected and attached to their local community thus increasing social support and hopefully decreasing the fears they experience. It is impossible to feel like an insider in a community of ever evolving change without a series of networks of other people. As the needs and wants of the older adults in Ferguson change, these series of networks will become increasingly crucial. Networks can be maintained through social integration in formal, more organized community groups or through an informal exchange of words with a neighbor across a dividing fence. Public casual conversation may seem trivial, but it establishes public respect, trust, and identity within the community (Jacobs 1961, p. 56).

"To create a 'sense of place' and foster connections among people, the physical characteristics of neighborhoods must draw people together and encourage an atmosphere of peace, security, and pride among residents of a community" and "while responsible participation in governance is important, getting to know your neighbors is a simple first step" (Roseland 2005, p. 156). Residents in Ferguson remember a time when they were more actively engaged with their neighbors. In initial conversations with a Ferguson resident prior to this research study, they suggested that neighbors do not stop and chat with each other anymore, that they did not really know their neighbors, and that they were more fearful because they would not even feel comfortable asking their neighbors for help if they needed it. It becomes difficult for older adults to trust their neighbors and feel comfortable talking about how the physical and social environment can foster safety and security. Especially when they do not feel like anyone takes the time to get to know one another. It may be that residents young and old are feeling similar frustrations with their sense of security, but have not taken the time to talk with their neighbors and community members about it. Collective action towards a common goal must first begin by discussing and deliberating about the issue and deciding what needs to be done, who will do it, and how a plan for action will be carried out. As with the older adults in Ferguson, "there is a longing, among millions of Americans now reaching middle age, for a stable and secure social world that they believe existed during their childhoods and does not exist now" (Dionne 1998, p. 93).

Although nostalgia may have both positive and negative aspects, it becomes one way to engage older adults in thinking about how to physically and socially improve their community to help reduce fear of crime. For older adults in Ferguson, nostalgia reflects a yearning for what they remember as a simpler life, full of comfortable and reliable relationships in their community (Dionne 1998, p. 93). "Residents of crime-ridden neighborhoods miss the street corner and doorstep sociability made possible by physical security" (Dionne 1998, p. 7). Older adults in Ferguson want back a neighborhood where they remember their children playing safely, where they could walk down the street safely, and one in which they did not live in fear of becoming a victim of criminal activity in their homes. They long for a place "where everyone knows your name, and probably a good many other things about you, and your commitments, and your family" (Dionne 1998, p. 3). They long for a network of community members capable of coming together for the collective good of the neighborhood.

References

Adams, R. E., & Serpe, R. T. (2000). Social integration, fear of crime, and life satisfaction. *Sociological Perspectives, 43*(4), 605–629.

Ames, D. L. (1995). Interpreting Post-World War II suburban landscapes as historic resources. In *Preserving the Recent Past, II* (pp. 97–101). Washington, D.C.: Historic Preservation Education Foundation. Retrieved from http://www.nps.gov/nr/publications/bulletins/suburbs/Ames.pdf

Appleyard, D. (1979). The environment as a social symbol: Within a theory of environmental action and perception. *Journal of the American Planning Association, 45*(2), 143–153.

Atkins, S., Husain, S., & Storey, A. (1991). *The influence of street lighting on crime and fear of crime.* London home office: Crime prevention unit paper no. 28. Retrieved from http://www.popcenter.org/library/scp/pdf/07-Atkins_Husain_Storey.pdf

Austin, D. M., Furr, L. A., & Spine, M. (2002). The effects of neighborhood conditions on perceptions of safety. *Journal of Criminal Justice, 30*, 417–427.
Bellair, P. E. (2006). Informal surveillance and street crime: A complex relationship. *Criminology, 38*(1), 137–170.
Bennett, G. G., et al. (2007). Safe to walk? Neighborhood safety and physical activity among public housing residents. *PLoS Medicine, 4*(10), 1599–1607.
Blazer, D. G. (2005). Depression and social support in later life: A clear but not obvious relationship. *Aging and Mental Health, 9*(6), 497–499.
Brown, B. B., Perkins, D. D., & Brown, G. (2004). Incivilities, place attachment and crime: Block and individual effects. *Journal of Environmental Psychology, 24*, 359–371.
Chadee, D., & Ditton, J. (2003). Are older people more afraid of crime? Revisiting Ferraro and LaGrange in Trinidad. *British Journal of Criminology, 43*, 417–433.
Christie, L. (2013). Most dangerous U.S. cities. *CNN Money*. Retrieved from http://money.cnn.com/gallery/real_estate/2013/01/23/dangerous-cities/2.html
City of Ferguson. (2014). *City history*. Retrieved from http://www.fergusoncity.com/141/City-History
Clarke, R. V. (1983). Situational crime prevention: Its theoretical basis and practical scope. *Crime and Justice, 4*, 225–256.
Coates, T. (2014, October). The racist housing policies that built Ferguson. *The Atlantic*. Retrieved from http://www.theatlantic.com/business/archive/2014/10/the-racist-housing-policies-that-built-ferguson/381595/
Cooperman, J. (2014, August). St. Louis: A city divided. *Aljazeera America*. Retrieved from http://america.aljazeera.com/articles/2014/8/18/st-louis-segregation.html
Cossman, J. S., & Rader, N. E. (2011). Fear of crime and personal vulnerability: Examining self-reported health. *Sociological Spectrum, 31*, 141–162.
Cozens, P. (2007). Public health and the potential benefits of crime prevention through environmental design. *NSW Public Health Bulletin, 18*, 11–12.
Cuba, L., & Hummon, D. M. (1993). A place to call home: Identification with dwelling, community, and region. *The Sociological Quarterly, 34*(1), 111–131.
Currier, J. (2011). Storm cleanup slowly progresses. *St. Louis Post-Dispatch*. Retrieved from http://www.stltoday.com/news/local/metro/storm-cleanup-slowly-progresses/article_13a7c4c5-b8d7-52d1-8504-25a12ad716a5.html
Dionne, E. J., Jr. (Ed.). (1998). *Community works: The revival of the civil society in America*. Washington, D.C.: Brooking Institution Press.
Duncan, M. (2008, Fall). Snapshot: An ordinary suburb, an extraordinary number of foreclosures. *Bridges*. Retrieved from https://www.stlouisfed.org/publications/br/articles/?id=504
Fox, T. (Ed.). (1995). *Where we live: A guide to St. Louis communities*. St. Louis, MO: Missouri Historical Society Press.
Gallagher, J. (2013, August). *North county suburbia in crisis*. St. Louis Post-Dispatch.
Gallagher, J. (2014, May). St. Louis is hotspot for underwater mortgages. *St. Louis Post-Dispatch*. Retrieved from http://www.stltoday.com/business/local/st-louis-is-hot-spot-for-underwater-mortgages/article_1a9b46b5-38f4-5b93-b4b4-e895f8e0bab5.html
Gans, H. J. (2013). *Violence and poverty: Images and realities of the underclass*. Retrieved from http://www.columbia.edu/cu/21stC/issue-1.2/Poverty.htm
Gay, M. (2014). White flight and White power in St. Louis. *Time*. Retrieved from http://time.com/3107729/michael-brown-shooting-ferguson-missouri-white-flight/
Gordon, C. (2008). *Mapping decline: St. Louis and the fate of the American city* (pp. 116–122). Philadelphia, PA: University of Pennsylvania Press.
Hanlon, B. (2009). A typology of inner-ring suburbs: Class, race, and ethnicity in U.S. suburbia. *City & Community, 8*(3), 221–246.
Hannon, R. (Ed.). (1986). *St. Louis: Its neighbors and neighborhoods, landmarks, and milestones*. St. Louis, MO: St. Louis Regional Commerce and Growth Association.
Herz, A., & Kania, H. (2002). Everyday perceptions of crime. *European Journal of Crime, 10*(4), 276–285.

Herzog, T. R., & Flynn-Smith, J. A. (2001). Preference and perceived danger as a function of the perceived curvature, length, and width of urban alleys. *Environment and Behavior, 33*(5), 653–666.

Hoffman, A. V. (2000). Why they built Pruitt-Igoe. In J. F. Bauman, R. Biles, & K. M. Szylvian (Eds.), *From tenements to the taylor homes: in search of an urban housing policy in twentieth-century America* (pp. 180–205). University Park, PA: The Pennsylvania State University Press.

Jacobs, J. (1961). *The death and life of great American cities.* New York, NY: Random House.

Joseph, J. (1997). Fear of crime among Black elderly. *Journal of Black Studies, 27*(5), 698–717.

Kelling, G. L., & Wilson, J. Q. (1982, March). Broken windows: The police and neighborhood safety. *Atlantic Monthly*, 29–38.

Kovarik, F. (2008, November). Mapping the divide: A lifelong St. Louisan grapples with the city's racial disparity. *St. Louis.* Retrieved from http://www.stlmag.com/Mapping-the-Divide/

Lancaster Community Safety Coalition. (2013). *Safety by design.* Retrieved from http://lancastercsc.org/our-programs/cpted/

Lawton, M. P. (1974). Social ecology and the health of older people. *The American Journal of Public Health, 64*(3), 257–260.

Lawton, M. P., & Nahemow, L. (1973). Environmental press model [Table]. Ecology and the Aging Process. In C. Eisdorfer & M. P. Lawton (Eds.), *The physiology of adult development and aging* (pp. 619–674). Washington D.C.: American Psychological Association

Lewyn, M. (2000). Suburban sprawl: Not just an environmental issue. *Marquette Law Review, 84* (301).

Lloyd, T. (2012, October). The battle over foreclosure mediation in St. Louis County. *St. Louis Public Radio.* Retrieved from http://news.stlpublicradio.org/post/battle-over-foreclosure-mediation-st-louis-county

Markson, E. W., & Hess, B. B. (1980). Older women in the city. *Signs: Journal of Women in Culture and Society,* 5(3), 127–141.

Miethe, T. D. (1995). Fear and withdrawal from urban life. *Annals of the American Academy of Political and Social Science, 539*, 14–27.

Minnery, J. R. (1986). *Crime perception and victimisation of inner city residents.* A Report for the Department of Planning and Landscape Architecture. Queensland Institute of Technology Brisbane, Australia. Retrieved from http://www.criminologyresearchcouncil.gov.au/reports/20-85.pdf

Missouri Department of Transportation [MODOT]. (2013). *Missouri's interstate system: yesterday, today, and tomorrow.* Retrieved from http://www.modot.org/interstate/MissourisInterstateHistory.htm

Morgan, P. (2010). Towards a developmental theory of place attachment. *Journal of Environmental Psychology, 30*, 11–22.

Nahemow, L., & Lawton, M. P. (1973). Toward an ecological theory on adaptation and aging. In *EDRA 4: Fourth International EDRA Conference* (pp. 24–32). Retrieved from http://www.edra.org/sites/default/files/publications/EDRA04-Nahemow-24-32_0.pdf

National Weather Service Weather Forecast Office. (2012). *Good Friday tornadoes April 22 2011.* Retrieved from http://www.crh.noaa.gov/lsx/?n=04_22_2011

National Weather Service Weather Forecast Office. (2013). *Severe thunderstorms produce straight line wind damage and nine tornadoes May 31 2013.* Retrieved from http://www.crh.noaa.gov/lsx/?n=05_31_2013

Newman, O. (1972). *Defensible space: Crime prevention through urban design.* New York, NY: Macmillan.

Office of Social and Economic Data Analysis. (2013). *Status of Seniors in St. Louis County.*

Pollack, L. M., & Patterson, A. H. (1980). Territoriality and fear of crime in elderly and nonelderly homeowners. *The Journal of Social Psychology, 111*, 119–129.

Quillian, L., & Pager, D. (2000). Black neighbors, higher crime? The role of racial stereotypes in evaluations of neighborhood crime (Working paper No. 2000-03). *Department of Sociology University of Wisconsin–Madison.* Retrieved from https://www.ssc.wisc.edu/cde/cdewp/2000-03.pdf
Relph, E. (1976). *Place and placelessness.* London: Pion.
Reynald, D. M., & Elffers, H. (2009). The future of Newman's defensible space theory: Linking defensible space and the routine activities of place. *European Journal of Criminology, 6*(1), 25–46.
Roseland, M. (2005). *Toward sustainable communities: Resources for citizens and their governments.* Gabriola Island, Canada: New Society Publishers.
Ross, C. E., & Jang, S. J. (2000). Neighborhood disorder, fear, and mistrust: The buffering role of social ties with neighbors. *American Journal of Community Psychology, 28*(4), 401–420.
Rothstein, R. (2014a). The making of Ferguson: Public policies at the root of its troubles. *Economic Policy Institute.* Retrieved from http://www.epi.org/publication/making-ferguson/
Rothstein, R. (2014b). The making of Ferguson. *American Prospect.* Retrieved from http://prospect.org/article/making-ferguson-how-decades-hostile-policy-created-powder-keg
Sampson, R. J., & Groves, W. B. (1989). Community structure and crime: Testing social disorganization theory. *American Journal of Sociology, 94*(4), 774–802.
Sampson, R. J., & Raudenbush, S. W. (2004). Seeing disorder: Neighborhood stigma and the social construction of "broken windows". *Social Psychology Quarterly, 67*(4), 319–342.
Scannell, L., & Gifford, R. (2010). Defining place attachment: A tripartite organizing framework. *Journal of Environmental Psychology, 30*(1), 1–10.
Schieman, S. (2009). Residential stability, neighborhood racial composition, and the subjective assessment of neighborhood problems among older adults. *Sociological Quarterly, 50*(4), 608–632.
Seamons, D., & Sowers, J. (2008). Place and placelessness, Edward Relph. In P. Hubbard, R. Kitchen, & G. Vallentine (Eds.). *Key texts in human geography.* London: SAGE. Retrieved from http://www.arch.ksu.edu/seamon/place_%26_placelessness_classic_texts.pdf
Skogan, W. G. (1990). *Disorder and decline: Crime and the spiral of decay in American neighborhoods.* Berkeley, CA: University of California Press.
Skogan, W. G. (1995). Crime and the racial fears of White Americans. *Annals of the American academy of political and social science: reactions to crime and violence, 539,* 59–71.
Taylor, R. B., & Covington, J. (1993). Community structural change and fear of crime. *Social Problems, 40*(3), 374–397.
Taylor, R. B., Shumaker, S. A., & Gottfredson, S. D. (1985). Neighborhood-level links between physical features and local sentiments: Deterioration, fear of crime and confidence. *Journal of Architectural and Planning Research, 2,* 261–275.
The City of St. Louis, Missouri. (2011). *A brief history of the City of St. Louis.* Retrieved from https://stlouis-mo.gov/visit-play/stlouis-history.cfm
Thompson, E. E., & Krause, N. (1998). Living alone and neighborhood characteristics as predictors of social support in late life. *Journal of Gerontology, 53*(6), S354–S364.
United States Census Bureau. (1990–2010). *American fact finder.* Retrieved from http://factfinder.census.gov/faces/tableserices/jsf/pages/productview.xhtml?src=CF, http://factfinder.census.gov/faces/tableservices/jsf/pages/productview.xhtml?src=CF, and http://censtats.census.gov/cgi-bin/pl94/pl94data.pl
United States Department of Justice. (2012). *Violent crime against the elderly reported by law enforcement in Michigan, 2005–2009.* Retrieved from http://www.bjs.gov/content/pub/ascii/vcerlem0509.txt
Wortley, R. (1996). Guilt, shame and situational crime prevention. In R. Homel (Ed.), *The politics and practice of situation crime prevention* (pp. 115–132). Monsey, NY: Criminal Justice Press. Retrieved from http://www.popcenter.org/library/crimeprevention/volume_05/06_wortley.pdf
Yin, P. P. (1980). Fear of crime among the elderly: Some issues and suggestions. *Social Problems, 27*(4), 492–504.

Chapter 3
Research Methods

Research Design

The purpose of this research study was to investigate older adults' perceptions of crime in Ferguson, Missouri and the impact on their quality of life. Data collection for this research study began in October 2013 and concluded in early June 2014. Transcendental Phenomenology was the qualitative method employed to understand the phenomenon. Purposive sampling, autophotography, photo-elicited in-depth semi-structured interviews, and a short survey were utilized for data collection. Ethical matters, the validity of the study, and the generalizability of the research results were also considered in relation to the design of the research study.

Description of Data Analysis

Transcendental Phenomenology

The transcendental phenomenological process varies slightly among researchers. The analysis methods utilized as part of this research study align more similarly with the methods outlined by Creswell (2007, p. 159) and as demonstrated through Moerer-Urdahl and Creswell (2004), providing a simplified version of the Stevick-Colaizzi-Keen method.

The following steps illustrate the methods outlined by Creswell (2007, p. 159) with the addition of more formal transcendental phenomenological language explained by Moustakas (1994):

- Epoche
  - Researcher is needed to suspend any presumptions he or she may possess about the experience under investigation (Conklin 2005).

K.L. Fisher-McLean, *Perceptions of Community Crime in Ferguson, MO*,
SpringerBriefs in Criminology, DOI 10.1007/978-3-319-46786-3_3

- Transcendental-Phenomenological Reduction
  - Horizontalization, delimited horizons, and clustering of horizons into themes
- Imaginative Variation
  - Vary meanings and perspectives of the phenomenon (Moustakas 1994, p. 180), describing *what* was experienced through textural descriptions and *how* it was experienced through structural descriptions (Moerer-Urdahl and Creswell 2004; Creswell 2007, p. 159).
- Synthesis
  - Formulation of a composite description of the phenomenon under investigation to arrive at the essence of the experience.

Sample and Population

The research study sample was composed of adults who were 60 years of age and older who lived within the geographical boundaries of Ferguson, Missouri. Traditionally, an older adult would be defined as someone who is age 65 and older. However, given the outcome of the pilot study conducted prior to this study, it was determined to lower the age criteria thereby increasing the potential sample size. Purposive sampling was utilized in this research study that involved recruiting participants who were within a particular predefined category (Trochim 2001, p. 56).

Each neighborhood within Ferguson has their own "neighborhood group" that meets regularly or semiregularly. An effort was made to attend as many of the neighborhood group meetings as possible to recruit research participants. Participants represented 8 of the 12 neighborhoods within the geographic boundaries of Ferguson, Missouri. These neighborhoods included Jeske Park, Old Ferguson West, Ferguson West (Wabash), Ferguson Hills, Forestwood, North Hills, Southwest, and Old Ferguson East.

An extensive effort was made to recruit older adults from the predominately Black Nesbit-Newton neighborhood located adjacent to the Canfield Green Apartment Complex. Residents of the neighborhood expressed concerns about crime and community disorder at their regular neighborhood watch meetings to police officers, the mayor, and the city manager. Unfortunately, no older adults from the neighborhood volunteered to participate in the research study. Some older adults from the neighborhood were concerned that their own personal safety might be jeopardized if other angry residents retaliated against them for taking photos as part of the research process. Even after providing them with the option to only participate in the interview portion of the research study, older adults declined. One Black male publicly expressed his frustration with the research study and felt that nothing would be done with the results to help improve the quality of their neighborhood.

Personally, it was felt there might be distrust with the researcher, as a young, White, female student from the university; a distrust that was probably warranted. Out of all of the neighborhood meetings in Ferguson, the researcher felt the most like an outsider in the Nesbit-Newton neighborhood. Research participation from this neighborhood was sought considerably because of the concerns with crime that they publicly expressed. However, there was a distrust that was going to be difficult to penetrate; a distrust that had roots extending deeper than the primary investigator's knowledge with the community or the problems with the neighborhood. These residents felt powerless in their repeated attempt to express their concerns to those who were in power; the police, mayor, city manager, and even the researcher. It seems that there may have been a long history of those in power not following through to address the concerns of residents and homeowners or at least the perception that those in power were not doing enough in the neighborhood.

Data Collection Procedure

Autophotography

Autophotography involves giving research participants a camera and asking them to photograph items of importance in their environment (Nolan 2006). In this study, research participants were supplied with a disposable camera by the primary investigator and asked to photograph physical elements and places that they felt symbolized crime in their neighborhood. Before participants were free to take photos, they were provided with brief verbal instructions on the picture taking process. Participants were instructed to only take photos of nonhuman physical objects and were instructed to keep such things as addresses and license plates out of pictures—elements that could potentially and specifically identify an individual. Participants were also instructed not to trespass onto any individual's property. All photos were to be taken from public view; this usually meant either the public street or sidewalk. These measures were taken to avoid any ethical or legal issues that might arise.

Photo-Elicitation

When photographs taken by participants are used as the basis for interviewing them, this process is known as photo-elicitation. An important element of photo-elicitation, or photo interviewing, is that it assists in developing a discussion in a more casual environment, taking pressure off the interviewee to be the "center of attention" (Tinkler 2013). Additionally, "the difference between interviews using images and text, and interviews using words alone lies in the ways we respond to these two forms of symbolic representation. This has a physical basis: the parts of the brain that process visual information are evolutionarily older than the parts that process verbal information. The images evoke deeper elements of human consciousness than do words; exchanges based on words alone utilize less of the brain's capacity than do exchanges in which the brain is processing images as well as words" (Harper 2002).

In addition to the photographs, a semi-structured interview guide, based on a state-of-the-art literature review, was also developed prior to the interviews. Some research participants completed taking the pictures in as little as a week, while some participants took a month or more to finish taking their pictures. Participants were not restricted on the amount of time they had to take the pictures. Although not specifically instructed to do so, some research participants also chose to takes notes about each photo they took. This information provided an excellent complement to the photographs and interview materials. Some research participants chose not to participate in taking photos, but still agreed to participate in the interview portion of the research study. A separate, but similar semi-structured interview guide, was developed for those who did not wish to take pictures.

Survey

A short paper survey was developed to collect demographic information. Participants were asked open-ended and multiple choice questions about their age, gender, race, the neighborhood they lived in, and whether or not they considered themselves an individual with a disability. It was felt that this information might impact how older adults perceived crime in their community.

Validity

According to Creswell (2009, pp. 190–193), the validity of qualitative research can be strengthened by applying a particular set of methods to arrive at precise results. This can be accomplished through the triangulation of various sources of data to rationalize the choice of thematic areas and by examining the accuracy of data by returning to the research participants to inspect particular accounts, and by expressing the results through the application of vivid descriptions to improve credibility. This is achieved through reducing researcher bias by clearing up and reflecting upon preconceptions, reporting upon a variety of perspectives (positive and negative) to add additional credibility, and researcher immersion in the research setting to develop a thorough comprehension of the phenomenon. Creswell (2009, p. 192) also suggests utilizing a person to examine the research study manuscript to assure that it also makes sense to an audience other than the researcher, and utilizing an external auditor not familiar with the research study to provide an objective assessment.

Photography in addition to the interview data provided various examples to support the themes present in the research results. Although two research participants chose not to take photographs, their verbal reports similarly aligned with those who chose to take photographs. Through the use of the photographs, research participants were also able to deliver vivid accounts of their experiences; pointing out particular examples to demonstrate their thoughts. The epoche process in transcendental phenomenology also helps to decrease researcher presumptions about the research topic, thus increasing the validity of the research study.

Due to continuous involvement in Ferguson, Missouri before and during the research study, full immersion into the research experience was achieved and thus made it more possible to fully understand the issues present in the community. Through the research results, differing viewpoints were accounted for in the explication of the thematic areas. Additionally, various external persons also reviewed the write-up of the research study to ensure that it would make sense to an outside audience. An external auditor, who also served as the final editor in the write-up, examined the write-up from the point of someone unfamiliar with research study. The combination of the above measures all helped to increase the validity of the research study.

Generalizability

Qualitative methodology is generally not "an appropriate means for arriving at statistical descriptions of a large population" (Babbie 2005, p. 343). Generalizability is utilized limitedly, because the results of a qualitative research study are not meant to be generalized to other sites or populations. Qualitative research studies tend to focus on the particular rather than the general in that themes are developed precisely for a certain population within a certain place at a particular moment in time (Creswell 2009, pp. 192–193). This research study investigated older adults' perceptions of crime in Ferguson, Missouri and should not be considered generalizable, neither to other older adults nor other communities. The results are not completely generalizable to all older adults in Ferguson, Missouri, as only 15 older adults volunteered to participate in the research study. However, the research methods employed are sufficiently generalizable to be applied to a similar population in other communities. This research provides a basis for a larger study that could be conducted utilizing a quantitative methodology that would provide more generalizable research results for Ferguson, Missouri as well as other potential communities.

Ethical Issues

Some older adults inquired about the safety of taking pictures related to crime throughout their neighborhood and the possibility of retaliation issues associated with taking these types of photos. Therefore, participants were strongly urged not to participate in the research study, if they did not feel safe doing so. They were reminded that participation in the research study was completely voluntary, and that if they chose to participate, but felt uncomfortable with the process at any point, they could withdrawal from the research study. They were also urged to avoid putting themselves in a position, where they felt unsafe taking a photograph. Research participants were also told that they could choose to only participate in the interview portion if they still wanted to contribute to the research study, but did not feel comfortable taking photographs. A local police officers spoke to some of the older adults and stated that he felt that the risk of retaliation by another individual, because they witnessed someone taking pictures in the community was quite low. He reassured older adults that as long as they stayed on public property, it was legal to take photos. Participants were instructed to only take photos of nonhuman physical objects and were instructed to try to keep such things as addresses and license plates out of pictures—elements that could specifically identify an individual.

Contact information for the participants was collected to deliver cameras and to schedule the photo-elicited interview session. This information was coded and stored on a password-protected computer. All additional information including photographs, interview data, and surveys were also coded so that the participants' identity would not be revealed and was kept inside a locked file cabinet inside the primary investigator's office. All research was approved by the University of Missouri Institutional Review Board.

Strengths and Challenges of the Research Study

Babbie (2005, pp. 342–343) recognizes that one of the strengths of qualitative research lies in the depth of understanding what the research study permits. The combination of autophotography and interviews with research participants provided an incredible depth of understanding about the phenomenon. The photos taken by research participants gave them a basis to talk about difficult issues in their community that they may have not otherwise spoken about in a traditional interview setting. The photos also were deemed useful in the manner in which they facilitated illustrating their experience in a way that words could not do on their own. The photographs also helped to provide tangible results outlining both older adults' concerns and points of optimism throughout their community.

Another strength of this research study was the representation of older adults. An extensive effort was made to recruit older adults from every neighborhood but was unfortunately unsuccessful. However, men and women, Black and White, varying in age from 60 to 78 volunteered to participate in the research study from eight neighborhoods in Ferguson, Missouri. The broad representation of older adults assisted in improving validity throughout the research process. The complete list of demographic characteristics is presented in Table 4.1 in the results chapter of this research study.

The research study also presented challenges. One of the main methods for recruitment included engaging older adults from various neighborhood and watch meetings held throughout the community. Therefore, it became likely that the research study would attract older adults who not only were interested in community crime, but who were already highly socially engaged. The literature review supports the notion that those who are more socially engaged, also tend to be less fearful of crime in their community. Those who are less socially engaged in their community and potentially isolated in a home environment might have a higher fear of crime. However, this is a population that becomes much more difficult to access for research study purposes.

Secondly, the research method of autophotography likely attracted physically mobile older adults. To assist in combating this issue, an opportunity was offered to older adults to only participate in the interview session of the research study. Two older adults chose to only participate in the interview portion. However, they were both rather active and physically mobile. Therefore, the modification of the research method as suggested above did not attract additional older adults to the research study who had mobility restrictions.

Another potential challenge was the time of year that the autophotography portion of the research study was conducted. Research recruitment occurred in the fall months, while the photographic portion of the research study took place in the late fall, winter, and early spring. Older adults were discouraged or delayed in taking pictures because of the harsh winter of 2013–2014 within the St. Louis area. A larger sample of older adults may have been recruited, if the photographic portion had occurred during the late spring, summer, and early fall months.

All older adults who participated in the research study were homeowners who lived in single family homes. This is a limitation because their perceptions about physical elements and places that they feel are representative of crime might be quite different from renters, those who live in privately owned condominiums, townhouses, duplexes, or apartment buildings. The sample of older adults, themselves were a limitation. Because the scope of this research study only analyzed older adult's perception of crime in Ferguson, Missouri, this cannot be considered an all-inclusive perception of what others may think about crime in Ferguson, Missouri. Those of other various ages may very likely have a very different perception about crime in Ferguson, Missouri. Although older adults were not asked about income as part of this research study, it is felt that most of the older adults who participated in this research study could be characterized as middle to upper middle class. This is a limitation because their perceptions of crime and their views about appropriate physical and social community norms may differ from those who represent a perceived lower class of residents, regardless of their race.

One major limitation was the researcher's perception as an outsider to those trying to be recruited to the research study. Although there was previous research engagement in the community, there were issues that presented themselves early in the recruitment of the research study. Several older adults expressed anger and animosity toward the research study. They deemed it as extraordinarily one-sided and having the potential to generate a negative image of their community. All concerns and criticism about the research study were taken extremely serious and as a result, modifications to the research study were made early in the process of data collection. In addition to photographing physical elements and places in their community that older adults felt contributed to crime, they were also asked to photograph physical elements and places that contributed positively to the image of the community. It was explained to community members that the primary investigator had a personal, professional, and ethical obligation to utilize research results for educational and community insight and improvement. The intent would never be to purposely cast a poor image of their community. The concern stems from a long standing reputation in which the local media has tended to focus on the negative aspects of North St. Louis County where Ferguson, Missouri is located; overlooking the positive aspects about the community. Those older adults who chose to participate in the research study were also pleased that they had the opportunity to photograph some positive elements about their community.

References

Babbie, E. R. (2005). *The basics of social research.* Belmont, CA: Thomson Wadsworth.

Conklin, T. A. (2005). Method or madness: Transcendental phenomenology as knowledge creator. *International Institute of Qualitative Inquiry.* http://www.iiqi.org/C4QI/httpdocs/qi2005/papers/concklin.pdf

Creswell, J. W. (2007). *Qualitative inquiry and research design: Choosing among five approaches* (2nd ed.). Thousand Oaks, CA: SAGE Publications Inc.

Creswell, J. W. (2009). *Research design: Qualitative, quantitative, and mixed methods approaches* (3rd ed.). Thousand Oaks, CA: SAGE Publications Inc.

Harper, D. (2002). Talking about pictures: A case for photo elicitation. *Visual Studies, 17*(1), 13–26. Retrieved from http://www.nyu.edu/classes/bkg/methods/harper.pdf

Moerer-Urdahl, T., & Creswell, J. (2004). Using transcendental phenomenology to explore the "ripple effect" in a leadership mentoring program. *International Journal of Qualitative Methods, 3*(2). Retrieved from http://www.ualberta.ca/~iiqm/backissues/3_2/pdf/moerer.pdf

Moustakas, C. (1994). *Phenomenological research methods.* Thousand Oaks, CA: SAGE Publications Inc.

Nolan, C. M. (2006). Auto-photography as research practice: Identity and self-esteem research. *Journal of Research Practice, 2*(1). Retrieved from http://files.eric.ed.gov/fulltext/EJ805685.pdf

Tinkler, P. (2013). *Using photography in social and historical research.* Thou sand Oaks, CA: SAGE Publication, Inc. Retrieved from http://www.sagepub.com/upm-data/53648_Tinkler_chapter_9.pdf

Trochim, W. M. (2001). *The research methods knowledge base.* Cincinnati, OH: Atomic Dog Publishing.

Chapter 4
Results

Population Sample

Fifteen older adults from eight of the twelve neighborhoods within the geographical boundaries of Ferguson, Missouri volunteered to participate in the research study between October 2013 and June 2014. Below is Table 4.1 which demographically describes each of the research participants and Fig. 4.1 illustrating maps of the Ferguson, MO.

Photo-Elicited Interviews

Thirteen of the fifteen research participants chose to contribute to the autophotography portion of the research study. Two research participants chose not to participate in the autophotography portion, but did contribute to the interview portion of the research study. Each individual interview was semi-structured and only the photos that individual took were used to guide his or her interview. Participants were interviewed at the location of their choice with most interviewed in their homes, while others were interviewed at other public locations in Ferguson. Through the transcendental phenomenological approach 12 themes emerged:

- community image
- interaction with nature
- undesirable attitudes and behaviors
- littering
- historic buildings and homes
- community safety
- community investment
- vacant property
- racial integration and community transformation

K.L. Fisher-McLean, *Perceptions of Community Crime in Ferguson, MO*,
SpringerBriefs in Criminology, DOI 10.1007/978-3-319-46786-3_4

Table 4.1 Participant demographic data

Participant	Gender	Age	Race	Neighborhood	Living situation	Disability
P1	Female	72	White	Jeske Park	Lives alone	No
P2	Male	63	White	North Hills	Lives alone	No
P3	Female	72	White	Jeske Park	Lives alone	No
P4	Female	71	White	Old Ferguson West	Lives w/spouse or partner	No
P5	Female	63	Black	Ferguson Hills	Lives w/spouse or partner	No
P6	Female	70	White	Old Ferguson East	Lives w/spouse or partner	Yes
P7	Male	74	White	Old Ferguson East	Lives w/spouse or partner	No
P8	Male	60	White	Jeske Park	Lives w/spouse or partner	No
P9	Female	63	Black	Southwest Ferguson	Lives with other in household (family, friends, roommate)	No
P10	Female	67	Black	Forestwood	Lives alone	No
P11	Female	61	White	Forestwood	Lives alone	No
P12	Male	65	Black	Wabash	Lives w/spouse or partner + grandchildren	No
P13	Female	72	White	Old Ferguson East	Lives alone	No
P14	Female	69	White	Old Ferguson East	Lives w/spouse or partner	No
P15	Female	78	Black	Ferguson Hills	Lives alone	No

- natural surveillance
- maintenance
- rental property.

Community Image

The theme community image developed as participants spoke about the displeasing aesthetics of some of the businesses in their community. They also felt that the presence of a particular population contributed to an undesirable appearance and affected the perception of safety in their community. However, participants also recognized that Ferguson had a number of community representations that were transforming that perception in a more positive direction.

As older adults spoke about the aesthetics of some local businesses that contributed undesirably to the image of their community, they utilized words and

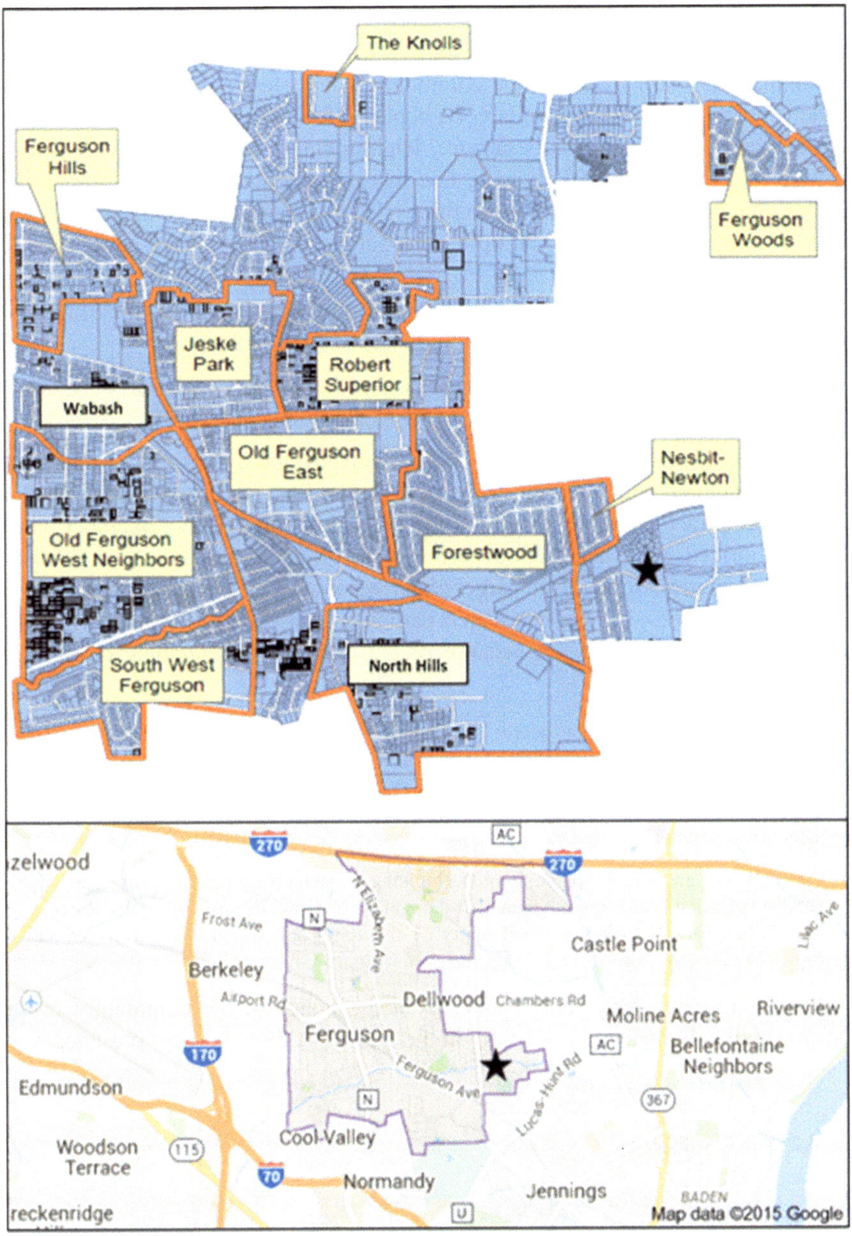

Fig. 4.1 *Top map* illustrates location of neighborhoods within Ferguson Missouri. City of Ferguson (2015). *Bottom map* illustrates location of Ferguson Missouri in relation to other surrounding suburbs and interstates. Retrieved from Google Maps, 2015. Ferguson (2015). *Asterisk* on each map indicates location of the Canfield Green Apartment Complex

phrases such as "cater to the poorer demographically and African Americans," "pathetic," "trashy," "indicative of the general deterioration," and "looks very inner city slum." One participant felt the community's homeless population contributed to the perception that "Ferguson is not a choice place to live" and "trashy down at the heels and possibly, probably unsafe." Highlighting an element that contributed positively to the image of Ferguson, one research participant noted, "sometimes North County does not get included in things that go on in the St. Louis region" and "a lot of the people who have an opinion about North County have never been to North County…"

One research participant spoke about her concern with the image of Ferguson in the context of planning for a class reunion,

> When I had been working on class reunions…so many of my old… schoolmates said 'oh no, I wouldn't come back to Ferguson. No, I don't go up there. It's not safe.'

Two research participants, both White females, both lifetime residents of Ferguson, captured what they felt were local businesses with displeasing aesthetics that did not attract desirable influences to their community. They remember when Ferguson thrived with successful department stores, drug stores, doctors, dentists, and other amenities. Some of the larger chain stores such as Walgreens have driven out small businesses, but these research participants also felt that some successful businesses had closed and left because of the dramatic demographic changes in Ferguson.

> …businesses that cater to the poorer demographically and African Americans…I really object to… their window displays. I don't like this image right there in downtown Ferguson…Businesses…look at the demographics…they aren't coming here…It makes me feel very sad…I know what this city used to be… It's just pathetic…indicative of the general deterioration… (Fig. 4.2).

> The chartreuse green in the painting on the building…looks very inner city slum to me … See that sandwich board sign out there?…I thought we had laws against having a sale every day of the year…It makes me feel angry…I think it just looks trashy as all get out (Fig. 4.3).

Another research participant spoke about the image of the community in the context of the homeless population in Ferguson. Although their presence was

Fig. 4.2 Window displays

Fig. 4.3 Chartreuse building color

intimidating at times, she did not feel that they were directly responsible for any sort of criminal mischief in Ferguson. More than anything, they just frightened people in Ferguson through their behavior and appearance.

> …a jeep…lived in by a homeless man who is mentally unstable. He's lived in Ferguson many many years…spends most of his time hanging out in the lobby of the library…It makes the gardeners uncomfortable…It kind of represents Ferguson's homeless people… It's a tragic problem…Sometimes through their behavior tend to make one uncomfortable…They sure don't look good. They are part of the perception that Ferguson is not a choice place to live…I think it adds to the perception that Ferguson is trashy down at the heels and possibly, probably unsafe… (Fig. 4.4).

One older adult White female chose to paint a more appealing picture of Ferguson, because she, like many other research participants felt that North St. Louis County was not portrayed optimistically in the media and not included in positive county-wide activities. As part of the celebration of the 250th anniversary of St. Louis, fiberglass cakes were placed in significant locations throughout the region.

Fig. 4.4 Jeep

Fig. 4.5 STL 250 cake

> …STL 250 presented the City of Ferguson with one of their cakes. They are 4 foot fiberglass cakes that have been decorated…Our cake is a fabulous representation of Ferguson …Sometimes North County does not get included in things that go on in the St. Louis region…A lot of the people who have an opinion about North County have never been in North County …They had a little presentation…folk drove by…people would honk and give us the thumbs up (Fig. 4.5).

Interaction with Nature

The theme interaction with nature emerged as both male and female, Black and White research participants spoke about the various parks and recreational activities throughout Ferguson that provided them with contact to natural elements. Research participants were very enthusiastic about the number of activities and progressive additions to parks throughout Ferguson, community gardens and farmer's markets, Ted Jones Bike Trail and other various events that the City of Ferguson hosted in an effort to engage residents and utilize their park spaces and neighborhood streets. One participant passionately spoke, "one of the best things about the City of Ferguson is our park system. You probably can't live more than a mile away from a park…I think our park system is a real asset to the community."

Older adults spoke about the various parks in Ferguson and used positive wording to describe their largest park, January Wabash as "beautiful," "peaceful," "quiet," "utopia," and "lively." However, one older adult felt that although the park was "beautiful," it could be very "dark" and "desolate," providing an opportunity for criminal activity. Participants also spoke encouragingly about the design and creation of new sculptures in Jeske Park and utilized wording and phrases such as "Ferguson is a happening place," "something of interest to all ages," "draw people into the city," and "interactive." One research participant reported that there was a creek running through the park that was being addressed by the city because in the past it had provided an "opportunity for crime" with "a lot of places where people would hide." The Ted Jones Trail, passing through Ferguson, was also a popular topic for older adults and was addressed as "positive," "progressive," "optimistic," and described as an opportunity for "freedom." However, it was also described as potentially "intimidating" depending on the time of day. Older adults also spoke about other various outdoor community activities including the farmer's market and Sunday Parkways event as an opportunity to "volunteer," "grassroots effort," "a gathering place," "positive," and something "for all ages."

One research participant spoke about outdoor interaction in the context of a regular event that symbolized the many community gatherings organized by the City of Ferguson and heavily supported by the volunteer efforts of community members.

> We have a Sunday Parkways…They had an event three times this summer… They close off streets…big tents go up…bike helmets are given out free…all kinds of activities…face painting…zoomba…A real grassroots effort by…people who volunteer. They don't go to the same park…each section of the city eventually has an opportunity to walk to that. Everyone city-wide is invited. They are always publicized…They draw a lot, a lot of families.

Several research participants spoke about their regular involvement in the weekly farmer's market and their support of the local neighborhood gardens in their community. Unfortunately, all of the photos taken by research participants were taken during the months when the outdoor farmer's market was not in regular session. Many participants expressed their disappointment in not being able to photograph the interaction of local farmers and community members at the market. However, older adults spoke about the market so vividly and with such excitement that it was easy to paint a mental image of the beauty of such an event.

> …the farmer's market is a gathering place. They shop for local produce… grown by the farmer who's standing there. There also is live music…food vendors…a spot to come see your neighbors…friends, meet new people…A real asset to the community…Something at the market for all ages…All the generations mixing together. Senior citizens are kept young by seeing kids doing chalk…hula-hooping…Makes me feel positive about what's going on in the community.

> We have two community gardens, east and west, where groups are growing vegetables… They meet a couple of nights a week and they share garden space. They are very hard working. They also provide food for the concert series that goes on all summer.

Research participants spoke about January Wabash Park and one participant explained the creation of the park by the Wabash Railroad Company for their employees to enjoy with their families. She also spoke in regard to the special memories she possessed about the park from her childhood. One occasion while visiting, a man caught a very large fish in the park's lake. She wanted to take a photo of the man with the fish for the local newspaper, but the man told her not to take the photo,

> This lake is such a gem, and I don't keep the fish, I catch and release. If you print a photo of how big this fish is, people will come and catch.

She respected the man's wish because she also shared the mutual belief that the park should be preserved for the personal enjoyment of the community, just as the workers of the Wabash Railroad Company enjoyed and as she had privately enjoyed throughout her life as a resident of Ferguson.

> I have never been up there and walked the trail when there weren't other people walking…I have never been there when there weren't fishermen. It's most beautiful in the snow… When you're on the three-quarter mile trail, when you get to the other end, it's peaceful and quiet… When you come back to…the lively parts of the park…the playground always has kids on it…You can hear young children laughing…See senior citizens fishing…It's a great gathering spot…It looks like utopia…A beautiful use of land with a lake and nature and ducks and geese (Fig. 4.6).

However, one research participant articulated that she felt leery about the spooky nature of January Wabash Park.

> It's beautiful, it's got a great walking path around it, but it's very desolate, very dark…I would not walk by myself there…It's a perfect spot to be robbed…Once you're back there you got nowhere to run from anybody…

She spoke about an incident in which a body was found wrapped up in the bed of a truck in the park and had concerns about the location of parking and the isolated nature of the footpath as it was wrapped around the backside of the lake.

Several older adults spoke about their fondness for Jeske Park, the addition of new interactive sculptures to the park, and the beautiful neighborhood that surrounds it. One research participant, who was a resident of the Jeske Park neighborhood, spoke about the addition of the sculpture park and the honor he felt in knowing one of the sculptors who grew up in the neighborhood. He hoped that the innovative enhancement to the neighborhood park would attract young families with children to their neighborhood. The participant also spoke about the excitement of a sculpture commemorating the past tornadoes. The local Christian Church had tongue and groove spruce beams that spanned the sanctuary. When the tornado came through, it went right over the church, throwing spruce logs up and down the street. The beams were salvaged by the sculptors and will be transformed into a wire and spruce twister commemorating the events of that day.

Fig. 4.6 Lake at January Wabash Park

Fig. 4.7 Jeske Park

> …it's certainly not the biggest park in Ferguson…It's going to draw people into the city… community…neighborhood when the sculpture park is finished. It's going to be a little different…I really do hope…it draws the 20 and 30 something's into look at the quality of the housing… Look at it as either a good starter home or someplace they could raise a family (Fig. 4.7).

> …the sculpture park will just say, wow, Ferguson is a happening place. There won't be any signs…do not touch, do not climb…interactive stuff…Introducing them to art…They are outdoors. They don't have a mobile device in their hands…Something of interest to all ages…

However, one older adult admitted that there were some opportunities for crime in the park that the city was addressing.

> There's a drainage ditch they laughingly call a creek that runs through here. But the city went in there and had the money to cut out all the honeysuckle, grapevines, and trees needing cutting…Actually it was an opportunity for crime because you're walking around that creek, there was a lot of places where people would hide. Kids would hide there and smoke and the things they wanted to do.

The Ted Jones Trail was also a popular discussion point among research participants and frequently utilized for walking and biking. One participant spoke of the Ted Jones Trail as being a gateway into Ferguson and the freedom he felt in riding his bike on the trail. However, he also acknowledged that there were parts of the trail that could be daunting. He described a tunnel that passed through the trail,

> It's so long and narrow, you can't see back in here. I used to go out in the morning when it is still dark and ride. And going through that tunnel—now you're not supposed to be in that park when the sun's not out. But going through that tunnel is, the thought, you could have just a flash of fear. Hey, someone could be over there waiting on me (Fig. 4.8).

Another research participant described the Ted Jones Trail in the context of her interactions with nature and animals on the trail.

> Shortly after getting on the trail last summer there was a young couple walking towards us. And the young woman said, 'be careful there's a snake in the middle of the trail.' I turned to

Fig. 4.8 Ted Jones Trail

them and said 'thanks for the warning' and then I ran over the snake with my bike. So I felt two bumps. I looked down and I literally freaked out. When we came back the snake was gone so apparently I didn't smoosh him because he slithered off the trail.

The trail was also described optimistically in the aspect of autonomous travel.

...it's a positive thing, it's an avenue for people to get around. If people wanted to live in Ferguson and commute to UMSL, it takes you right into the campus...I've travelled it a lot...a freedom in getting out and riding my bike...I don't know of any reported crimes on it since it's been open...I see it as a progressive thing ...An optimistic thing... This part of this is probably the loneliest part of it as you're approaching going up into Cool Valley and if you're coming down out of Normandy up from Florissant Road.

That's not graffiti. It's artwork that has been posted on there and it's kind of graffiti like. So, don't look at that photograph and think that that's incriminating. That's not (Fig. 4.9).

Fig. 4.9 Artwork on tunnel

Undesirable Attitudes and Behaviors

Several older adults were concerned about some of the generalized behavior exhibited by various individuals in their community. Particularly they seemed to be concerned with behaviors and deterioration of values amongst youth in the community. Words and phrases that older adults utilized to specifically demonstrate some of the issues they observed among community youth included, "fighting," "cussing," "screaming," and exhibited behaviors that are "intimidating to seniors," "irritating," and "unwanted to the elderly." Older adults also felt that some of the undesirable behaviors such as "hanging out," and other "annoying behavior" displayed by various community members demonstrated an attitude of "no values," "not caring," "no self-respect," "rudeness," and a "lack of civility."

A Black male spoke about the undesirable behaviors he witnessed as neighborhood children walked home from school.

> ...overall a lot of riffraff...more confusion going on...children coming from school fighting, arguing, cussing...parents...not instilling the correct values in their children.

Another older adult also felt that children were no longer held accountable for their adverse actions and they had no fear that there would be any negative implications for their behaviors.

> ...I wonder, too, about the kids that walk around...Why aren't they in school?... When I was a kid...all the police knew you and they knew your mom and dad...If you were out on the street when you weren't supposed to be, you were in trouble. That just doesn't happen now.

There was also some concern amongst both Black and White research participants about youth walking in the street, disrupting traffic, putting themselves in harm's way, and intimidating the general public through their actions.

> I personally am not comfortable when I see groups of African American youths walking up and down the middle of the street...They're always screaming at one another or fighting.

> The groups...intimidating to seniors...walking in the streets instead of on the sidewalk.

In particular, one research participant spoke about the frustrating behavior of youth playing basketball in the street near her home. This attracted undesirable individuals to the neighborhood and expressed unpredictable behavior that was unwanted by homeowners in the neighborhood.

> ...the basketball goal...It would bring unwanted traffic...Not only the homeowners or the children...they bring their friends from other neighboring communities...The code enforcers should put something up on the property owner's door...to tell them that they have to move that...People drive down the street...real fast...they're out there in the street...It's just kinda irritating...I wouldn't want it next door to my house...It's just not safe for them...It's unwanted to the elderly around here. I would remove that basketball goal and encourage them to utilize the park area...They should know not to have this...going towards the street...They could hit the cars and cause property damage...It's just more of a nuisance... (Fig. 4.10).

Fig. 4.10 Basketball goal

A Black male communicated his concern with the values of some of the people in his neighborhood, their fear of reporting suspicious activity, and their lack of involvement in keeping the neighborhood safe.

> Certain people in the community have this…no snitch philosophy…if I saw somebody going in your house…garage…back yard…taking something that I know didn't belong to them…I'm quite naturally going to tell you…Give you a description of who did that…I don't understand…when I was growing up…everybody stuck together. You better not do anything wrong…somebody would go tell your parents…No values…not knowing your neighbor…not caring…not respecting your neighbor…not respecting your neighbor's property…

The same research participant felt that many of the undesirable attitudes and behaviors exhibited by young adults boiled down to a lack of self-respect and lack of motivation to change their lives. He describes returning from the Vietnam War, feeling some of the same social pressures that these young adults do, and the process of making positive changes in his life.

> …you see the people…the way they dress…the way they act…I can look at myself… what I used to do…when I was in Vietnam…especially young folks are doing, I did all that… When I came back here, I knew I couldn't do that…I said to myself, 'what I'll do is I'll

> leave that there…come back and use the time that I have to dry out'…They haven't reached a point in their lives where…they want to change…It's no self-respect, don't care…A general negative attitude.

Two White females were also uncomfortable with the general offensive behaviors exhibited by individuals in public settings; particularly in retail stores while they were trying to do their shopping.

> …there are times of day that I will not go to the grocery store, the drug store…The people I see make me feel uncomfortable…Their idea of personal space is different than mine… Annoying behavior like the king of the aisle deal…like nobody else in the store exists…Not aware of their responsibilities of being one of many.

> I do not like it when I'm at a stop light and hear…very loud music booming from the car next to me…Having to be super alert when I go to the grocery store…General rudeness and lack of civility exhibited.

Finally, one research participant seemed to be concerned with individuals of working age who seemed to be hanging out on their front porch; day and night.

> …we've got areas where there are a lot of…undesirable people who hang around the front yard and the porch…I would not be able to say, 'well, those are the people that would commit crimes'…When there's a lot of hanging around…they can be up to no good sometimes…

Littering

The theme littering developed as White and Black, male and female research participants spoke about the disturbing amount of trash present throughout their community. Trashed seemed to include anything from a small amount of waste such as candy wrappers and soft drink cans, to grass clippings, and larger items such as tires and automobile bumpers. One neighborhood was described as "makes you think about the ghetto," to describe the problem with disrespectful illegal dumping. Littering was also described as "a growing problem," and "ignorant." This made older adults feel "hopeless," and that the situation was "discouraging."

One older adult described the vacant lot where a neighborhood used to be present across the street from her home. The airport bought up the land, tore down the neighborhood and all of the streets, then sold the land to a development company. The participant felt that not as much dumping occurred when there was still a neighborhood presence. Since the land has been vacant, people utilize the area to dump trash, assuming no one is taking ownership of the land or that someone else will clean it up.

> …somebody in the middle of the night thought that I needed a used tire…left it right in front of the house…That just blows my mind…It means you are living in an undesirable area… You'd rather look out…see the grass cut, manicured nice and clean…It makes me angry really, that someone would throw their trash in front of your house…I wouldn't take my trash and throw it in front of theirs (Fig. 4.11).

Fig. 4.11 Abandoned tire

...bags of trash, leaves...branches...car parts; bumpers...They just put it in their nice clean truck...drive it over...dump it off...It makes you think about the ghetto, not being comfortable in an area where people are just trying to survive. Who lives here? Who would want to live here?

I've called the police because people are dumping stuff...They have no regards for people living there... They tend to do it on vacant land. They think, well the airport will clean it up...The airport don't live here.

The research participant also described an area down the street from her house, across from the vacant land. There was a large amount of unbridled growth blocking residents' view of the vacant land; disrupting the natural surveillance they might otherwise have. She also described the contagious nature of trash dumping, almost as if unspoken permission was being given to others who were thinking about committing a similar action.

Just paper... trash...This is the side that people dump the most...because the residents can't see...If they would clear it out that would stop...They would think people are watching... They wouldn't be dumping stuff... (Fig. 4.12)

The hardest thing is to keep it cleaned up. If you dump something, get it cleaned up right away so that other people don't think, 'oh, we can bring our stuff and dump it there.'

Another research participant described a creek used for dumping that lacked any opportunity for natural surveillance and/or territorial reinforcement over the area.

Fig. 4.12 Unbridled growth and trash

> …by Maline Creek. It appears to have already been used for a drinking and dumping area…beer cans… a tire here by the culvert… Whoever is responsible for the creek access by the road has not taken precautions …to minimize…trash dumping…Maybe a fence should have gone up…Somebody's taking advantage of this area. So how do we minimize that from occurring (Fig. 4.13)?

Several older adults spoke about their own personal pursuit to keep their community clean. They wanted to create the image that someone cares about their home, neighborhood, and community. One White female described how she and her friends were disappointed in how a particular area of their town looked, so they took it upon themselves to put together a weekly group of volunteers to pick up trash along a busy thoroughfare. Unfortunately, she also spoke about the disappointment that they felt in their volunteer efforts.

> …a liquor store…that disturbs people…They feel it attracts an element that is not healthy for a community. A group of friends and I used to go up every Sunday morning and pick up litter…We finally stopped… It was so hopeless …After about 3 or 4 years …we quit… discouraging (Fig. 4.14).

Another research participant spoke about doing her part to keep the neighborhood clean by picking up after her dog when she walked with him.

> I have in both of my cars…at the office, poop bags. We do not let our dog poop and we don't clean it up…I want my neighborhood to be representative of me…It shows respect.

Fig. 4.13 Maline Creek area

Fig. 4.14 Liquor store

One research participant and her husband walk regularly for exercise and on their routes they pick up trash in an effort to beautify their community.

> …My husband and I are big about littering. He is going to be known eventually as the 'trash man cometh'…We walk daily…We have different routes…As we walk he picks up trash…It is totally amazing what people will throw out of the cars or leave in the parking lots…

Two White female research participants spoke about their disappointment with young people littering in the neighborhoods on their way home from school in the afternoon.

> …they send little food packages home with children…They eat the fruit… drink the juice and dump the cans…I picked up a large bag full of litter as an example…The neighborhood has definitely changed…A growing problem, but believe me people my age did not litter. Your parents would have dealt with you if you had…They used to say about Ferguson was that if you did something bad on the other side of town your parents knew about it before you got home…Just total disrespect…It's a cultural thing people didn't see any problem with this trash… Where do you think the term trashy neighborhood as an insult came up. It's because trash is a problem. It degrades the neighborhood…You need a place to look as if someone is claiming it as their area…They will not tolerate that here (Fig. 4.15).

> …it is such a throw-away society…Not only a racial situation, but an age-related situation. I see kids of all races pitching stuff, but I tend to see it more with minorities…That bothers me…In the time I grew up there weren't as many disposable containers…sacks…papers… My parents were children of the depression so they were very frugal…It just makes me sad that we don't have more respect not only for our own property but for other people's property…It sends a message that people don't care…not conscientious…I don't think that that's necessarily a true representation, but I can see how somebody passing through might draw that conclusion.

One Black female described her place of business and how she regularly exercises both natural surveillance and territorial reinforcement to minimize dumping on her parking lot. She describes a car rental place across the street and how people were using her empty lot to clean out their cars before returning their rental vehicle.

Fig. 4.15 Litter

> …There's a car dealer shop…People returning their car…He must have told them you're going to have to have the car cleaned out…They literally clean their car out on my lot. They don't know that I can see them…They can't see in…The way I turn the blinds…The first time I saw them do it I said, 'oh, look how ignorant'…The second time, 'it's not happening for you my brother…'

By taking control of her area, she also explained how a lot of dumping has stopped. Her persistent effort to both keep an eye on people's negligent behavior and the actions she exhibit around her property show that someone cares for and is claiming the property.

> …a lot of dumping has stopped…People see a clean area and they are less likely to litter… When it's busy traffic, that's when I clean my windows. It's when I clean my doors. I'm out there sweeping…I want people to know…

This participant also described how there are several trash barrels up and down the street near her place of business and how that has helped to eliminate excuses people might have to dispose of their trash in the street.

> …this is a group effort…As many people as we have walking, you would think this place would be filthy…We are picking stuff up. There are…beautiful brick trash barrels…You've got stuff in your hand…if you're not being stupid, then why would you throw it on the ground?

Finally, one White female suggested that despite her discouragement with littering in her community, she had faith in the younger generations as their schools take them out and encourage them to pick up trash around their local parks.

> I see hope…January Wabash Park, students from the middle school periodically go around and clean up things that have been thrown into the lake…It is a good lesson that we are at least making people aware that this is not the way you want to do things.

Historic Buildings and Homes

The theme historic buildings and homes was developed as research participants emphasized the value that the community placed on preserving the historical traditions and sites throughout the City of Ferguson. Research participants utilized words and phrases such as "makes me happy," "all well taken care of," and a "positive thing" for the community. One participant wholeheartedly affirmed that the historic school, once attended by her grandmother, made her feel "nostalgic." Another older adult felt that Ferguson should be proud of its heritage of stately homes, but needed "to do a better job of promoting ourselves."

One research participant described how Ferguson originated as a train town and the community-wide effort to preserve the historic train cabooses in which families frequently visited.

Fig. 4.16 Train cabooses

> …a train town…How Ferguson came to be… Business people got together… purchased two cabooses. The Ferguson Historical Society…rehabbed them… authentic on the inside… open for tours… draws in tiny little kids to senior citizens …It makes me happy… I like to see people interacting with things that are around town…I love the historic part… Driving down the main thoroughfare of the street… you see a couple of red cabooses, you just think, 'wow, that's cool (Fig. 4.16).

Research participants felt that the historic homes throughout Ferguson were particularly significant. They believed Ferguson had historic homes that rivaled anything found in what were considered more desirable parts of the county. They also showed a great interest in ensuring that these homes were cherished and preserved for future generations.

> …It is classic; gorgeous craftsman, big eaves, big porches on both sides… An example of a beautiful home…We have never been a braggadocios community. We don't try to sell our selves…We need to do a better job of promoting ourselves…This home looks like it's loved…It says we are thrilled to be here…We are willing to take up upkeep on our lawn, the house itself… Someone who is proud of their home…I've only lived in old homes. I have a real soft spot for charm (Fig. 4.17).

Fig. 4.17 Example of historic home

...This could be in Midwest Magazine. I don't know that these are the images that people would have in their minds that they would find in North County.

One research participant chose to take a picture of her own home to demonstrate her affection for older homes and the sense of attachment she experienced for not only her home and memories inside, but the neighborhood that surrounds it. She explains that when any updates or repairs need to be made to the home, she always considers preserving the original architectural integrity.

This is my home ...raised 3 children here. It was built in 1908...There's one backyard separating this from the home I grew up...a great place to raise kids. They could wander the neighborhood as I did...always playmates around...grew up knowing their neighbors...We remain there today...with knees that are 69 years old and laundry that's in a basement...My comfort zone is very small. I've only moved two homes away from where I grew up... we're probably there to stay...Look at that front porch...There's a porch swing on there 12 months of the year...antique wicker furniture on there all summer...baskets hanging, strawberry crates, artwork...It looks like an outdoor living room. I like to see people walking their dogs...kids two doors away riding by on their bikes...It's just a window on the world...Everyone on the street is friendly and everyone does watch out for one another...That's part of being a community. I don't want to live in a house and not know who's around me...I'm really proud of the fact that you can live in a home for 48 years and still be happy; that it would have served your needs from the time you were raising your family until you were a senior citizen...It's been good to us (Fig. 4.18).

She also captured the home that she grew up in, close to the home where she currently resides. The photo serves as a descriptor of some of the beautiful historical homes in her community, but also as a basis to reminisce about the memories she spent with friends and family.

...this is where I grew up...It's a lovely home...Any child who was raised in a home like this would have just a warm and fuzzy feeling. We could entertain extended family. We had a huge, very deep lot...A great bar that I climbed up on when I was a kid...swings back there...a barbeque pit...family picnics...A home made for entertaining...it makes me feel really young. My girlfriends and I would sit out there. Boyfriends and I would sit out there after dates. It's just a beautiful home and a beautiful setting (Fig. 4.19).

Fig. 4.18 Research participant's home

Fig. 4.19 Research participant childhood home

Another older adult demonstrated the pride of ownership exhibited by residents despite issues occurring within close proximity of their home.

> …this street abuts to some houses that are not in as good repair …These people…take good care of their homes…They are not fearful of what's two or three blocks away …There has been some crime over in that section…These people hang in there…All Victorian homes… all well taken care of…They care about the community…They care about their home… about the image that shows to people that come into town…A positive thing…an indication of their character… (Fig. 4.20).

The oldest school in Ferguson also symbolized a historical anchor and a personal anchor for an older adult progressing through life in the community.

Fig. 4.20 Example of well-maintained home in Ferguson

...one of the oldest schools in Ferguson...This means stable firm community. Where people would want to raise their kids...It's a beautiful neighborhood...I think my grandmother went there...It's an attraction to the city...nostalgic...

Community Safety

The theme community safety developed as participants communicated about the various aspects of crime and safety in the community. Overall, male and female, Black and White older adults communicated, "we really feel safe," "I feel comfortable," "I don't feel afraid," and "people are looking out for the residents." However, some also agreed that "opportunists see older people as potential victims," and that it was important to be "cautious about my surroundings." One research participant felt that "it only takes a handful to ruin a community," and another states that her "doors are locked at all times." Two research participants spoke about the criminal activity present at some of the local apartment complexes, stating that there had "been a lot of troubles in the past," but at the time of data collection (late 2013–mid-2014) felt like tensions had been alleviated. One older adult felt that most of the crime in the apartment areas was "self-predatory," an "unfortunate situation", in which the victims must feel "powerlessness." Finally, some older adults spoke about direct victimization or victimization they witnessed to property or other people in their community. They described their personal feelings as, "I felt violated," "I felt frustrated," and "sad."

Several residents felt that Ferguson itself was not the main problem when it came to crime and violence in the community. Both Black and White participants voiced concern with crime in the context of nearby neighbors that brought unwanted chaos to the community.

One of Ferguson's main problems...are our neighbors – Berkelely, Dellwood, Cool Valley, and beyond. These areas have a lot more crime...but quite a bit of traffic through our city.

...in Berkeley...known as the drug strip...a lot of gang activity...Since they had a major bust...I haven't heard any gun shots...since the summer...That's pretty good...One day...I heard some guns shots...from an automatic pistol...I went out...and see if I could find these cartridges...Put them in a plastic bag...give them to the police...It angers me... They're either fighting over turf or drug debt...Don't own any property in the neighborhood... Not paying any taxes...Don't have a job...Bringing down the community...with the thought...they can get something for nothing...It only takes a handful to ruin a community...

Two female research participants, one Black and one White expressed their concern with security bars positioned on residential doors throughout their neighborhoods.

...security doors... that means ...there are a lot of break-ins ...on a regular basis... There are other ways...that you could secure your house; better lighting, alarm systems...The down and dirt way is to stick up bars...Most of that kind of crime happens within a mile of where the person lives...That means they're with us. They're our neighbors (Fig. 4.21).

Fig. 4.21 Bars on door

The majority of the research participants stated that they felt very secure in Ferguson and did not restrict their day-to-day activities around the community.

> We really feel safe in our neighborhood…There's a new business…We were talking with them about why they chose here…They did not have to worry about where they walked at night…

However, some also acknowledged that crooks preyed on the feeble mind and body; targeting those who appeared to be an effortless mark. Therefore, one was wise to not foster conditions that would brand themselves as a suitable target.

> …opportunists see older people as potential victims…When my mother got to a certain age there was no way…I would let her go up to the grocery store by herself…tottering…If I was a young thug I'd see easy pickins'. Dangling a purse…the whole bit.

> …predators…look for the weak. The lion looks for the wilder beast that is kind of stumbling…As you get older, that certainly is a factor. It doesn't necessarily have to be an extenuating circumstance…Predators go where there is easy pickings…deep pockets. We've got some friends that live over on Ladue Road…their house got broken into. They moved out of North County to get there…if you're gonna be a crook you go where the money is…it would be wise to be aware of your circumstances and not look to be a victim…Most criminals are cowards. They take the path of least resistance…don't be the path of least resistance.

Some research participants, mainly females, acknowledged that their spouses, friends, and family were more concerned about their safety than they were. Personally, they were reassured that the individual procedures they adopted assisted in their defense.

> …this is my community…I don't worry if I walk home from a city council meeting in the dark …if I walk home from a friend's…This is my town…You don't want to mess with me…My husband will say, 'I wish you would be more aware…I wish you would be more careful'…I belong here…I don't walk around meek and mild, like there is always a boogey

> man. I walk around like I own this sidewalk. I'm not walking haltingly. I look like I can take care of myself.I just take my cell phone, my stick…I'm very conscious of the area…I just walk a different way…each time…If someone is watching me, they don't know which way I am going…I don't feel afraid. I feel happy. I wave to people and speak to people…I don't want to be worried. It interferes with your activities…I haven't been afraid since I got robbed…when my youngest daughter left…I said…'I am not going to be afraid.'

In an extreme endeavor to ensure the safety of his wife in her late night adventures, one husband sent their daughter to look after her as she strolled through the neighborhood. The Black female also conveyed the gratitude she had for neighbors who watched over her like an extension of the family.

> I went out for a walk about 10:30…The cop got out of the car and they pulled my daughter over…I said, 'what's the problem?'…They said, 'Oh, we got a couple of calls…they said somebody was stalking you'…People in my neighborhood know that I walk late at night… they called the police…But what happened, my husband sent my daughter to trail me…so I didn't know she was trailing me. It was my own family…I asked her, 'what are you doing?' She said 'dad told me to come follow you'…Wasn't it a beautiful thing that I have neighbor?

Older adults expressed their sense of safety in different manners. One research participant articulated that safety meant that she and her dog could roam freely without reservation.

> It makes me feel warm and fuzzy that I have my dog…This is not my dog. I was babysitting this dog for five days, two years ago…Here's an animal that can't do anything for himself. Dependent totally on me… He can roam around and enjoy…I don't have to be afraid that somebody's going to steal him…I've made a safe haven even for the dog (Fig. 4.22).

She also had no personal intention of altering any of her current activities, walking proud and strolling through her community as she always had.

> …out walking the dog….Lots of young people walking around, mostly guys…Their pants hanging down…Some of them knew me, some of them didn't…Even though its busy, I'm still able to take my dog out with no reservations…The guys are walking slow…I'll say, 'coming through'…They open up and let me through…That says a lot about my community.

> Most people think when you see a liquor store there's all the people hanging out…People would…shy away from there…I'm not shying away from there. I'm going up there.

Fig. 4.22 Dog

> I'm passing there…I got a lot of respect from those people that stop in there…A little corner store for a lot of people that don't have cars …It's so orderly and clean that I don't have a problem passing it.

A Black male communicated his frustrating experiences as a young paperboy in North St. Louis City and observing older adults taken advantage of and robbed regularly. He also acknowledged the extreme security measures some older adults have taken in order to protect themselves.

> …younger thugs have a tendency to feel that they can take advantage of older people…In the neighborhood that I grew up in they…were less reluctant to try to rob…older people… They were afraid they might be armed…Not only that, people respected older people…

Male and female, Black and White research participants spoke about their satisfaction with the work that their local police officers were undertaking to maintain safety in the community; expressing that police genuinely cared for and were concerned with the safety of older adults. One Black female recognized that police regularly passed by her place of business to provide her assurance that they were watching out for her.

> Safety. I've got a lot of windows…I do have an alarm system…I am at my building sometimes as late as 9:00 and 10:00 o'clock…I am not a bit afraid. The mayor has let me know that he will have policemen coming by…Every now and then I will hear a horn blowing…It's a policemen…just passing through…It gives me assurance…People are looking out for the residents.

She also spoke about the safety that she felt in her neighborhood where her home was located and how the police regularly patrol that area.

> …in the morning the doors are open. All down my block the doors are opened…I can see policemen coming up and down my street two…three times a day…I feel comfortable there.

However, not all of the older adults were as optimistic about their personal safety in the community as others were. One White female stated:

> I no longer feel comfortable working in my front yard…I never leave my car parked outside…I have an alarm system (I couldn't continue to live here by myself without it). My doors are locked at all times.

Another Black female stated that although she felt very safe during that day in her community she recognized that there were certain threats that existed during the evening hours.

> I don't travel a lot at night…I'm very cautious about my surroundings…I park my car in the garage…I have to come out of the garage to get into my front door…I would be afraid that somebody can come up on me especially if they know I live alone…

Two White males recognized that although most of Ferguson was relatively secure, there were particular areas where more crime appeared to transpire.

> There are areas that are a part of the City of Ferguson that are probably more problematic… Apartment areas…across West Florissant that used to be called Canfield Green…There have been lot of troubles in the past, but recently hasn't…There's the Sharon Dale Apartments behind McDonalds over by West Florissant, but they're putting a lot of money into. They put up wrought iron, six or eight foot tall fences all around the property…It's nice looking…I would feel safer…The fences would prevent someone from parking along…Ferguson Avenue…running into the apartment building…I see that as some real improvement to those apartment areas.

> By the transient nature of the apartment buildings that we have…There's robbery… shootings…domestic violence…It's unfortunate…It's primarily on the victims…who live there…It's self-predatory…An unfortunate situation…The city can send our cops in…I think the most frustrating thing is…it must be terrible to be in a mindset that is so hopeless that you allow that to continue…It seems like…people involved with that would say enough of this. But it doesn't occur…There must be powerlessness that goes along with that. I don't know how to image that.

Research participants spoke about particular measures taken by the City of Ferguson to enhance the safety of their community.

> This is our new emergency sign…There's five of them…If you're walking…see something happening there's a phone…It automatically go to the 911 operator…Anyone that would see this would say well the city's really taking a step to protect their citizens…but it depends on how you look at it. So there must be a lot of crime here; look they've got these on every corner.

A couple of research participants spoke about reckless driving being a major issue in the community and some of the measures that the City of Ferguson had taken to try to address the problem.

> The speed on my street…The city has tried to address…Put the dividers in the street…They paid for a study…because we had people coming down that street at 60-70 miles per hour.

Research participants spoke about specific crime-related incidents in their community and the impression it had left upon them. One Black female communicated how she felt after discovering damaged business vans that she utilizes for her public charity work.

> …one of my vans was vandalized…The same night, they removed a catalytic converter off of one of the other vans…I felt violated. I felt anger…I'm there a bunch of hours to give support to those in need…The cops stopped by…They felt sorry for me…When you've been violated…you think about it all the time…I started using the lights…bought the burglar alarm…I felt frustrated…I thank God that I was not angry at what we do…It never crossed my mind to quit…It made me more determined…We are doing goodwill…But I do want you to know…this is not utopia.

She also spoke about an incident in which the local veterinarian office was robbed and the frustration and perceptions associated with the broken window on the front façade of their building that have not yet been repaired.

> …there's a window here boarded…That bothers me…Out of all the goodness that we have going on, there are some things that can be distracting. This looks like there is some type of…decay. When you see boarded up buildings, you wonder why…When people don't have the right information, the mind can conjure up anything …Could have been a

Fig. 4.23 Business burglary

shooting…a robbery. Somebody could have died…they had a break in…They had their monitors working… had their alarm set… somebody was bold enough to go in and take lab taps…dog heaters…That caused me pause…I have been in my building at 3:00 o'clock in the morning… (Fig. 4.23).

One White female spoke specifically about the number of incidents that she had experienced in her own home. These included two separate break-ins and several windows shot at with pellet guns throughout the years.

…we have had quite a number of incidents in this house…We put that lock in because at 3 o'clock on a Sunday morning a man tried to kick in the door so forcefully that the door is permanently warped…Our back door is a replacement…that got kicked in…That has really messed up my ability to sleep through the night.

…there's my little pellet gun. This is in my garage…Why haven't I gotten it fixed?…I'm waiting for it to happen again…I also have a little pellet…in one of these side light storm windows…From the same kid…he shot out quite a number of the other windows in our barn which is older than the house…It was a real loss and it was too expensive for us to replace it with that same quality glass…His punishment…I taught him how to replace the screen…

One research participant spoke about the sadness that she felt when she learned that someone's property in her neighborhood was vandalized. Although she felt sad, she also believed this was a very isolated occurrence.

This is a house down the street…In front the homeowner…planted a row of small trees… Somebody came through…cut down all the trees…They certainly weren't mature trees… they weren't little saplings…We stopped one day…I said how sorry we were…It sounded like it might have been targeted for personal reasons…I don't think it is indicative of something going on in the community…it just made me sad because I hate to see trees destroyed…We have just lost a hundred-year-old tree in our front yard…I was particularly feeling bad…

Another Black female described her own past personal experience being robbed at gunpoint, but also described how she has not let that affect her future endeavors.

I was robbed…at the Normandy bowling alley…at gun point…I had my keys in my teeth… I was picking up my ball to put in the car…he said, 'just give me the purse, just give me the purse.' I handed him the purse…I went back into the bowling alley and told the people that I had been robbed…The police, they caught him…I had to be in my 50's…I didn't want to go bowling anymore alone…I was not going anywhere alone anymore at night time…I've

> gotten over that...The only thing that's keeping me from going out at night is my vision. My vision is very, very bad. If they don't pick me up, I don't go.

Finally, a White female described an event that occurred shortly before her research interview in which a young lady was robbed at gunpoint outside of one of the local bars in Ferguson. Precautions were now taken to minimize victimization in the future.

> ...7:30 on a Saturday evening. She was getting in her car...He pulled his car around and stopped behind hers and said he had a gun and...said 'give me your purse'...she tried to... talk him out of it and he said 'lady I'm not messing around' and pulled the gun...she pushed his hand away...he grabbed her and pulled her out of the car and took the purse and left...I thought nothing of tripping on out to my car...Now you're more aware of how you get in your car; you pull your door shut quickly...I look to see is there any one standing around me; are there any cars that are just sitting?...I hurry up get in my car and lock the door. The guys go out until we get our cars started and drive away and then they come in.

Community Investment

For the older adults in Ferguson, community investment depicted an immeasurable phenomenon. They reported inspiring stories of people investing in their community and in turn the community reinvesting in them as a maturing adult. They also spoke about their encouragement in witnessing young and vibrant visionary investors taking a leap of faith in their community to make it a better place. This also included the impressive efforts of the City of Ferguson to improve opportunities and infrastructure to promote engagement among their residents.

Older adults described Ferguson as a "hidden jewel" with many "positive things going on." They felt "fortunate," "hopeful," and "optimistic" about the amount of positive "progress" occurring in their community. Older adults also spoke of their "inspiration" in the number of volunteers in their community working toward the "betterment of Ferguson" and how countless volunteers came together, assisting friends, family, neighbors, and strangers, with tornado damage repairs from past years.

Research participants exhibited an almost indescribable affection for their neighborhood and community. Neighbors had quickly become friends and then family through the years. One older adult explained that being part of a community meant,

> loving where you are, loving it with all its quirks. It's hard to explain the feeling...From the very beginning we had felt like we belonged here.

When one is part of a genuine family, they become invested and are able to look beyond the minor blemishes and idiosyncrasies that exist, demonstrating an authentic affection unfounded anywhere else.

Older adults spoke about how their love for the appearance of the "house" drew them into the neighborhood and community. However, it was the bond that developed and moved them beyond a beautiful and suitable house, to a home with neighbors they trusted. Neighborhood children became their nearest and dearest, just as neighbors had fostered a relationship with their own children when they were small.

> …we bought this house…We liked the house…When we found out that we loved the community…we were…beside ourselves…We were part of…history because this…is an older community.

Some research participants grew up in Ferguson, raised their children in Ferguson, and now entertained their grandchildren in Ferguson. They are invested in Ferguson. Ferguson is their home, big or small.

> …neighborhoods are people…I love it. They think…when you have arrived, you should have a big house. I'm very comfortable…When I want to come to a big house, I'll come to your house.

> I haven't moved more than two yards away in 69 years…I'm either very blessed or very boring….I'm always told that I am very blessed.

Another lifetime resident, so indefinably acquainted with her community, could not imagine living anywhere else. She was invested in her home and neighborhood while she cared for her aging mother.

> I like it. It's comfortable. I like my house. It's quite. It's pretty. I like my yard. My house is perfectly comfortable for me…It's convenient to everything I know…My mom and dad live within blocks, my dad is deceased now but my mother still lives there…I visit her all the time.

One Black female spoke about her neighborhood in comparison to the hustle and bustle of city life that she had come accustomed to; she had become invested in Ferguson's tranquility.

> …overall I still enjoy my neighborhood…When I first moved here I couldn't even sleep at night…It was so quiet and peaceful. I was used to the noise.

One older adult invested in her community by conveying a helpful and approachable manner to those traveling through her community in need of assistance. She described a circumstance in which a man driving through town was trying to find his way to a particular street. He asked another man on the street for directions, but he was unable to assist him.

> There was a new guy that I never seen when I was out on this day…He was trying to get to Georgia…I said…'you have to go back to Harvey…make a right…then you will be on Georgia'…It never crossed his mind to approach me…What…I did…was let him know… in this neighborhood we are approachable… Guess what, the old lady knows how to get you where you want to go.

Ferguson has been plagued with a series of misfortunes over the past few years. However, research participants spoke about how some of those tragedies have unified their community and brought them closer together.

> Community—when you get those first four letters…put them to the side…put a dash after the last "m" you have the word 'unity.' That's what we are. We are unified.

> …it is a neighborhood…There are people that care about one another…A neighborhood is first a place where I live. This is my dwelling place. This is where I rise and shine.

Research participants spoke about their experience with the tornadoes that occurred in 2011 and 2013.

> I think people tend to help each other. When we had that tornado, people in the streets and people helping out the next day. I don't think you find that everywhere.

> All the churches came together…We worked like dogs…FEMA estimated that we did over a half a million dollars' worth of work for that first tornado…The CERTs…People could not afford, with their insurance, to get that kind of work done…We had people that came from Iowa…come from everywhere with CERT…chain saws…big as me…We came together…It was beautiful…We had a lot of men on my team and I was the one, a woman here, that's in charge. They had no problem…they called me 'ever ready' because everything we needed was ready…We came together…We had meetings before and after. It was just wonderful…A lot of us got to know one another…We was working so close together…

> …Urban Chestnut, you ever been down there? It's a nice restaurant downtown. I heard the guy that is the manager…on the radio…He said that they got all new furniture made from the oak trees that came down…from the tornado two years ago.

Male and female, Black and White older adults volunteered heavily in other daily activities and events throughout the community.

> There's a group in Ferguson that heavily volunteers…You get sucked into it.Do not label me a racist – I am a realist. The racists have already moved…I am among a very dedicated group…who work very hard in various volunteer capacities toward the betterment of Ferguson. There are many wonderful residents, business owners…city employees… Ferguson is my home, I love Ferguson and hope I can always live here until I too, get 'carried out feet first.'

> …I service people…in need…Pack up food for them at night…in the mornings the food is ready to go…It's a big turn over…I get a lot of emergencies…It's not like I can tell somebody, 'oh, come next week'…We're total volunteer…nobody gets paid…More people are in need than ever before…During the day I'm talking on the phone…ministering… prayer…They are looking for hope.

Research participants also invested in their community through their participation in other various events throughout the community and the surrounding area.

> Everything that goes on in Ferguson…the neighborhood meetings, I go to that…birthday parties around our little ten house block…I love it…We trade off barbequing…At Thanksgiving and Christmas…they'll come over and get dressing…then they'll send me a crumble cake…

> My friends, we all hang out…the little bar across the street…If we got out during the week, we just grab…hamburgers…chopped salads…They have dart contests…All kinds of stuff going on…

…group of our friends…We meet at the local bar one night a week…We're all involved in the community as far as different committees…We like to BBQ…enjoy each other company…I just like being involved…I like the outcome…It's satisfaction when you're done.

I swim at the Y…bowl at the Big Webber…I go dancing…go out to the clubs…I walk a lot…I've been an active person all my life…I'm still doing my own lawn.

We visit the restaurants…attend the farmer's market…use the walking/hiking trail that goes from downtown, the Ted Jones Trail over through UMSL.

…there are a lot of people that do a lot of walking in Ferguson…As the weather starts to break in March, you'll see a lot of people riding bikes…I never seen anybody getting mugged. I never heard of anybody being harassed…Not in the 30 years I've been there.

Several older adults spoke that they had invested in the small town accessible feel of their community. It was a place that was safe for their children to walk to school on their own and a place where they felt comfortable growing old.

…we both came from smaller towns…It feels like a small town…in a big metropolitan area.

We were talking to an exchange student in Florissant…She was saying that she felt so isolated…She was from someplace in Europe…She could walk every place or bicycle… here you had to have a car…That really struck home…I thought…'that's not true of Ferguson'…I can walk to the post office…grocery store…library…restaurants…That enhances the feeling of community…As we get older and contemplate not being able to drive, that is also a plus.

Male and female, Black and White research participants had consciously invested in their neighborhoods by getting to know everyone in their block.

I like that neighbors know one another. I know everybody on my block. I know everybody on the block in front of me….Even though my house butts up into another subdivision… know everybody on that block…it's the camaraderie of the neighborhood…

It always feels safe. Maybe it's because we have made the efforts…We have laid roots…. We met the neighbors because we ran and jogged, so we would see people and stop and talk…

Know your neighbor…That's easy to do…That is contagious. When you grow up greeting people that you don't know…other people see you do that…

I describe Ferguson as a community with people who live in houses with front porches… greeting the people…walking the streets…There are… communities where you pull into your subdivision…pull the garage door up…pull the car up…You don't know your neighbor…Speak to your neighbors…socialize with neighbors…This is an asset…that draws people together…

People need to get to know each other, to interact…exchange phone numbers…ideas… Once the residents show an interest, the city shows more of an interest…They still function off that squeaky wheel gets to oil thing…If you are not in their face telling them about it all the time, then you kind of get left behind.

Several older adults, both Black and White, spoke about how important they felt it was that young families with children invested in their community; to enjoy the same lifelong fulfilling relationships that they had enjoyed.

Fig. 4.24 Girls walking home from school

> I'm always glad to see young people in our community. It's extremely important to me that families live in Ferguson. If my husband and I as senior citizens leave, it's not such a big deal. But you always need those next generations...to see grade school kids coming and going to school. I have a garden in the front yard. The kids always speak to me. They will ask me what a plant is. If they feel comfortable walking the street, I feel comfortable having them out there.
>
> ...a group of girls who were walking home from Blessed Teresa of Calcutta Parochial School...It sort of typifies to me what living in the old town part of Ferguson is...in this day and age that's really important...The girls after school bonding... That's a real draw for families to be able to move somewhere where their kids can walk to school (Fig. 4.24).
>
> Because of the size of the houses...price of the houses, we are seeing a lot of young families moving in, which is refreshing...very hopeful...There are a good many people in the neighborhood that realize the value of the neighborhood and the community...

Not only was it important to see young families with children, but they also expressed their desire to be a positive influence on young people in their neighborhood.

> ...I would get out here...try to learn how to...talk and deal with the younger generation. To give them some inspiration and some hope...I used to tell my son, and my daughter...'it may seem like I'm raising...a lot of hell, but...I want you to do better than I did...if you and your sister did better than I did...I've done some correct'...that's all I want...

One Black female spoke about teaching young people to be respectful neighbors and living harmoniously with one another.

> The little guy on the corner, he's been there for about nine months...I heard this 'baboom, baboom, noise'...I went over in my robe and...said, 'hey, this is a neighborhood. You can't have this noise here...I don't want to have to put a sign up about noise.' 'Oh, no ma'am, no ma'am.' I said 'okay, tell all your friends'...He told all of his friends don't come here with all that noise...They work in a band...They are real good, too...but you're not going to play over here like that.

She spoke about how they have developed a mutual respect in their neighborhood and how the young man comes over and borrows things from them, seeks advice, and asks for help with various tasks.

> …when he first got his lawn mower, he came over…asked my husband to show him what to do…He bought this big barbeque grill…he called my husband over and said 'can you come over and show me how to use it?'…When he had to go somewhere where he had to wear a tie, he came over…and showed him how to tie the tie…

All older adults were very enthusiastic about the continuous growth and improvements they saw in their community. They were excited about investors coming to the area and the time and money that the City of Ferguson was investing to improve resources and opportunities for their citizens and visitors. Several of the research participants were also heavily invested in their community through various opportunities.

Nearly every research participant spoke about all of the positive progress occurring in Ferguson. They were particularly excited about the new community center being built. Participants were enthusiastic about the prospect of having a place to hold meetings and community gatherings. They spoke eagerly about the grass roots efforts that many of them were involved in to launch the project.

> Ferguson has a ton of projects going on right now…a new fire department …police department… community center…A number of good, positive things going on. If you've always lived here…you are very thankful…People…have lunch from Emerson or Express Scripts and they say, 'wow, we never crossed the property line coming into Ferguson. We didn't know this was here.' This is a great hidden jewel…that's important…it says we care. We are moving forward…progress.

> …when you have a community, you need a community center…We have a number of youth organizations here…they need somewhere to gather…I thank the city staff… city council for realizing that this was something that our community was lacking…They are willing to put the money into this…As people drive by they are thinking, 'wow, look what's going on in Ferguson (Fig. 4.25).'

Research participants boasted about the community-wide effort to assist in obtaining the church to transform it into the new community center.

> …the citizens got together and said save that church…It is one of the cornerstones of our community…It took door-to-door canvassing…asking for votes…The archdiocese said they would sell it…With the money that the city paid, with the increase in taxes, the church was able to upgrade…we got the property.

Fig. 4.25 New community center

Research participants also spoke about how fortunate they felt having a major university close to their hometown. They felt that having a major university nearby not only provided opportunities for community members that might not otherwise exist, but that various business opportunities, students, staff, and professors were drawn into Ferguson because of the proximity to the University of Missouri—St. Louis.

> …Mark Twain Recreation Center at UMSL …The city is really fortunate to have the university…Classes that are open to the public…a new pool…They are good about letting …the surrounding community use their facilities…We have a number of apartments …for university students…If they are living there, they are also eating there…joining in community events…When their experience is positive, they are telling people about that… (Fig. 4.26).

In addition to the new community center, older adults also felt that the City of Ferguson was doing a number of other things to promote community engagement throughout the city. Many of these community engagement opportunities were free to residents in an effort to promote socialization.

> …they have concerts there in the summer …fall, spring. It's a place for people to gather… listen to music. It's maybe two or three years old…It's a good feeling…It looks like it's something the city did to…promote community togetherness…The city is concerned about…older retired people…It is something that they can do that doesn't cost them anything and get enjoyment out of it (Fig. 4.27).

Older adults in Ferguson were thrilled about the number of positive projects developed by investors in their community. In the past few years, several stores, restaurants, apartments, and other businesses have been revitalizing the main strip in the old town part of Ferguson.

> …a new young man… has taken this building…added apartments above it… This is what can be done …if someone has faith in the community…Puts a smile on my face. I'm always encouraged when someone has the ability…money… time…to start a big project … Another great building with opportunity to bring in new residents…new business people on South Florissant Road (Fig. 4.28).

Fig. 4.26 UMSL mark twain recreational center

Fig. 4.27 Community gathering place

Fig. 4.28 New downtown development 1

The unofficial community center in Ferguson has become the Corner Coffee House. Several research participants spoke about their amazement in the rehab of this building, how it has become a meeting place for several community organizations, a place to meet friends, and grab a cup of coffee.

> …the Corner Coffee House…was an auto parts store…before that it was a car dealer …He took this extremely filthy, disgusting, dirty, greasy building… made it into a charming coffee house …There are teachers who stop there every morning for coffee… senior citizens who gather there…Most of them are men…If your group has a meeting, you can hold it down there…Been our community center for a number of years…Makes me smile to see all these men sitting around communicating…They are all cheerful …laughing…teasing one another…There is somewhere you can go in the morning…you know that a friendly face is going to greet you…It also shows that…a small community-minded man can start a business like this and have it be successful (Fig. 4.29).

Older adults also expressed their excitement with a number of other businesses that have invested in Ferguson over the past several years. Several research participants told their own individual stories about a progressive minded lifelong resident who decided to invest in making his home a better place. He invested not only architecturally and in terms of community development, but in the soul of Ferguson. Everyone felt individually touched by the man's efforts to improve the city. These positive contributions regularly overshadowed crime-related events that the media chose to cover about their community.

Fig. 4.29 Corner Coffee House

...Things have changed a lot. When I was growing up...there was brick buildings all along the street...pictures of Ferguson in the 50's, 60's...almost all of those buildings are gone... For a long time there wasn't any new stuff going on... This building was the beginning... He did this...he did the Brew House...his son has The Cork and Vincenzo's...Now he has had a stroke...his son has taken over...Now we have a new investor in Ferguson...There's a lot going on... (Fig. 4.30).

...the Post-Dispatch is writing up good reviews about the restaurants ...There's an element here that you find in some of the more trendy suburban cities; Cottleville, Webster, Kirkwood...When people start talking now about urban blossoming, they are talking about places like Richmond Heights...Ferguson has come into....That kind of counters the murder...domestic violence calls that you hear about on the evening news...The totality of this is optimistic (Fig. 4.31).

One White male spoke about the optimism that he felt about new investors revitalizing some of the historic homes in the area.

...it's a home...It's vacant...They are repairing it...You feel hopeful when you see this going on... People are interested enough to try to preserve what we've got...the oldest frame home in Ferguson...A beautiful house ...They sort of talked about it being a party house for wedding receptions...It shows a lot of pride in the city...Someone who had the money to do it stepped up to preserve it...North County...is thought of as not having very much to show...It certainly is a show place. People had talked about this house through the years...You feel good that someone recognized the value of it... (Fig. 4.32).

Fig. 4.30 Ferguon Brew House

Fig. 4.31 New downtown development 2

Fig. 4.32 Remodeled historic home

Vacant Property

Male and female, Black and White older adults were particularly concerned with the amount of vacant homes in their community. Some of those homes were vacant due to foreclosure and some were due to tornado damage that occurred in 2011 and 2013. Vacant properties made older adults anxious about decreasing property values in the neighborhood, the image of the neighborhood, and crimes that may occur in or around those vacant properties. They described various vacant properties throughout their community as a "genuine mess," a "black hole," "an eyesore for the neighborhood," "extremely dangerous looking," and "tacky." Research participants believed that these properties "cast a bad light on Ferguson" and they expressed "frustration with the city" because "no progress seems to happen" on some of the properties. They also felt that the situation with the enormous number of vacant homes was "frustrating," "unfortunate," "upsetting," "discouraging," and "disheartening."

One older adult was concerned about the appearance of some of the vacant homes in her neighborhood and the length of time that the foreclosure process seemed to take.

> I saw a couple of houses that are really looking kind of bad…I know there is one down the street…where the grass is growing quite tall…sometimes the door is left open. It's probably in the process of being foreclosed. The foreclosure process just drags on and on…I think it had a blue tarp on it at one time…I don't know if they corrected that problem or not. But then there's another house…It's vacant and the gutters are falling down…It's looking very tacky.

Unfortunately, in conjunction with the lengthy foreclosure process the essential components of the home begin to fail as it sits uninhabited. Historic community gems become quickly dilapidated and frequently if repaired, cheaper materials are used, compromising the architectural integrity of the home.

> We have two vacant houses…just past my next door neighbor's house, they have both just been sold…One has been vacant for a couple of years…The people just walked away… Then the one right next to this one was a darling little red brick…Tudor style…on a double lot…It had a beautiful tile roof…it did have a leak. So did Bank of America fix the leak? No, they ripped off all the tile roof…put on a cheap asbestos roof…there goes half the character of the house.

Another common concern among research participants were the official notices placed on windows and doors on the front of the house. These notices become an obvious sign that no one lives in the house and thus became a target for criminal activity.

> The house next to us has…official notices…Every month they send somebody out to check on the house…I commented that that is a real problem…that says…'boy, here I am empty now'…legally they had to do that…We are very watchful of the house, but we are not here all the time…

Several research participants commented that they had vacant properties next to or near their own property in which they had personally taken it upon themselves to partially maintain or keep a watchful eye. Two male research participants mowed the lawn at neighboring vacant homes to make it look lived in and less of an attraction from crime. This also helped to maintain the image of their home and property. One research participant spoke about his frustration with a man breaking into the vacant house next door to him and the action that he took to help prevent it in the future. These residents have become the unofficial neighborhood watch group, looking out for and reporting suspicious activity in the community.

> …this house was vacant…The guy I see him going through the neighborhood all the time…I saw him in the back of the house…I immediately called the police on him…When he saw me he left…He had removed the garbage disposal out of the house…I went over there…put some screws in the door…so that he couldn't walk go back in…He would have to break a window…I'd automatically hear him…When I called the police I told the officer…'that a personal threat to me because if the guy comes over here on my property…

climbs over that fence…touches anything… I'm going to take person action against him. You won't have to, I'll do it.'

People in the community were particularly upset about the condition of this home because it is on a main thoroughfare in the community and is considered one of the historic charms. It was well documented by several research participants.

This is the old century house…Some opportunistic flippers got in and they either bit off more than they can chew or they just don't care…It's just a genuine mess…They bought the house and then the tornado happened…The one we call the Good Friday…One of the things that Ferguson is proud of is the old Victorian houses. The old railroad… houses that were built; two-story, three-story…this was one of them…Some of the gems of the community don't get the attention that you would think would…On my bad days, it shows the neglect of the city…There are a good number of people that look at some of those houses …and say those are the prime of the city…It just casts a bad light on Ferguson as a whole… If this were in Kirkwood, this wouldn't occur. And there are houses in Ferguson that rival anything in Webster and Kirkwood (Fig. 4.33).

Another common theme were homes that were abandoned or turned into rental property because of older adults passing away in the community. The homes were not worth enough money to encourage their family to do anything and are left to deteriorate and become a blight in the neighborhood.

This is a house that has been empty for 10 years…It overlooks Jeske Park… An elderly gentleman lived here died. All the family is out of state…can never come to a conclusion on what should be done with the property…People have looked at it and wanted to buy it. The city has contacted them. They do not keep it up. It's not for sale. They have refused any overtures…It's a black hole…It is an eyesore for the neighborhood…It's a level of frustration with…housing values in the city…If it was a quarter of a million dollar house sitting there…somebody would have been doing something with it…There's days that I feel wonderful about living in Ferguson and then there's days that I'm thinking, 'man, I've got to get out of here quick… (Fig. 4.34)'.

Other older adults expressed a tremendous emotional attachment to vacant and neglected homes in their communities. Their friends and families once resided in

Fig. 4.33 Old century home

Fig. 4.34 Abandoned home overlooking Jeske Park

these homes and they had very fond memories from when they were growing up. In some cases, research participants almost appeared to give these homes humanistic qualities and characteristics as they described the heartache they suffered over their declining state.

> …I went to brownie meetings in this house when I was a little girl and one of my friends lived there…makes me sick…How could you misuse a house and a property that badly?… The city should never let it get that bad. They seem to go around Ferguson with their little blinders on and pass things…Aren't these inspectors going by and seeing these things on their own? Do we have to be complainers and tell them?…It's just frustrating…breaks my heart…such a waste… The owner doesn't care, the city doesn't care enough to protect everybody on the block (Fig. 4.35).

Fig. 4.35 Abandoned brownie meeting home

Fig. 4.36 Abandoned high school best friend's home

...there's nobody that lives here. These bushes are all growing up...their bird bath... It's knocked over... no windows...no curtains ...It just looks like... somebody could see no one lives here and... steal copper...It's unfortunate... My best friend in high school used to live in this house...It means something to me...I've been in this house...We would walk home from school every day in junior high and high school...and we would watch American Bandstand. Iconical...I hate to see it treated like this (Fig. 4.36).

How many of those in the neighborhood do you need? ...The general idea is...things are not looking good...My neighborhood is maybe not as healthy as it used to be.

One research participant felt discouraged about the manner in which society had begun to view homes in the community as merely disposable properties.

...a vacant home... boarded up...an abandoned property that could be a health hazard, a fire hazard, people could move in there...Kids could go in there...do drugs... My tendency like most peoples is avoidance ...It's upsetting ...discouraging...disheartening, it makes you feel as though any money that you sink into your home now is just an absolute waste; you're just flushing it...it's...difficult to tell what is neighborhood deterioration and what is social deterioration...Thinking of...your home...as strictly financially just an investment... was a very bad move for society to take...It has hurt communities...made it just seem so horribly self-centered ...Took away a lot of the feelings that people ought to have about their homes and communities (Figs. 4.37 and 4.38).

Research participants expressed their frustration with homes that had unfortunately been damaged by fire. These homes were left vacant in a dilapidated state for many years with no action taken on them.

...this is a house that caught fire...almost two year ago...nothing has been done. I don't know whether they are waiting on insurance...It is so discouraging...discouraging for people who live around it...It certainly has to affect property values...I've been in the house...It's a good house...We have another one in the neighborhood that caught fire about a year ago...They just sit there...I don't know whether it is an attraction for crime, but... doesn't speak well for the neighborhood.

...I was...in the neighborhood that day the house caught fire...something really needs to be done here because all the houses around there are impeccably clean...This means that the property value in that neighborhood is going to go down...It upsets me that that was such a

Fig. 4.37 Abandoned, boarded-up home

Fig. 4.38 Dilapidated home

nice neighborhood…One that I grew up in…I knew the people that owned the house originally, it was a gorgeous house …boarded up…just sitting there waiting for something to happen…A boarded up house, literally with boards looks a whole lot worse than just a vacant house…

Boarded-up homes were more disturbing to research participants than homes that were simply vacant. Boards were most likely present as a security measure to keep vagrants from busting out all of the windows in the home or to cover up windows and doors that had already suffered damage. However, research participants agreed that a building covered in boards conveyed a grimmer reality than did a merely uninhabited house.

…vacant houses that are boarded up that appear vacant…What I mean by that is if you have a vacant house… you leave the drapes up, leave it look like its lived in…keep the grass cut, people really don't know its vacant….If you put plywood over the windows…over the doors so that people can't come in and steal your copper well that looks like an invitation… They know nobody could be living in a house that's got that kind of stuff on it (Fig. 4.39).

One research participant thought that if the situation necessitated that a vacant home must have plywood over the windows and doors, that it would be beneficial if

Fig. 4.39 Abandoned home; plywood over doors and windows

the plywood were painted the same color as the house as to not make it so obvious that the home is empty.

> …you've got some big tall bushes by the front door… They can hide someone entering this vacant house. The plywood over the windows is painted to match the house so you don't see the issues right away …because the wood is painted…There was some effort to make it not become an eyesore in the neighborhood… (Fig. 4.40).

> …you can't tell if the car maybe belongs to the neighbor who parked it in the driveway to make it look occupied or if someone in the house is there. But because of the board up on the window it makes you think that it's vacant. The car, don't know who it belongs to makes you think…maybe it might not be vacant…The neighbors trying to…reduce the vulnerability of this house to burglars…I'm not sure if there's a mailbox there or not…I worry…if I see a property that doesn't have a mailbox at the curb… somebody's knocked it out and nobody's put it back up…Therefore, it indicated to whoever knocked it out that there's an opportunity available (Fig. 4.41).

Vacant homes in the community where nothing was done to improve their current condition often conveyed to older adults that the City of Ferguson did not care about the deteriorating image of the community and the effect it had on other individual's property in the neighborhood.

Fig. 4.40 Abandoned home; painted plywood

Fig. 4.41 Abandoned home; neighbor parked in driveway

> …there could be crime…They could be using that house…It's getting more… deteriorated and the people on both sides of this house keep a very nice yard…That's gotta bring down property values…That's a perfect place for people if they are selling drugs…They could be…inhabiting that house…trash all over the front yard, the doors hanging off. Would you want to live next door to that?…The neighborhood's deteriorating and the city…doesn't care…They wouldn't have let that go that long…I go by it every day going to work…Every day's worse and worse.

> …this has been this way for years…The tornado's hit the roof…They have not done anything about it…The city has not been vigilant about…vacant properties, run down properties. There have been others in the last ten plus years that the city has purchased… torn down…rehabbed…This is one that they have not touched. Part of it is anger that it's been allowed to sit there and nothing be done…It shows deterioration…That the city doesn't really care…

In some cases, research participants were happy to see progress on vacant and dilapidated properties in their community.

> …big equipment bulldozing a house down. This house was damaged in the tornado… beyond repair…This was something positive to me…in this instance, the homeowner and the city did not let that linger. It was taken care of… There's progress being made… It no longer drags the rest of the homes on this street back. It shows progress…It makes me feel like someone cares (Fig. 4.42).

Several older adults were also concerned about the condition of a number of vacant and distressed businesses in their community. Two White research participants, one male and one female mentioned the condition of the Math Center.

> Math center. It used to be a bowling alley. Then it was a batting cage area…then it sat vacant for 10 years… then it was donated…but it's just a shell…There wasn't anything in it but just mold…There's a roof on it but here is leakage…The city continually harassed her about it while there are other properties like that one…When she was still here a year ago she called me to tell me that someone had gotten into her stuff and stolen some of the things…What she found …was a room…on the back of one side…all her stuff was back in there…Some kids had taken some tables and apparently gotten three fold out tables…An

Fig. 4.42 Bulldozed home

old cash register that was of some value…was in there…That has since been sold…She lost her house through default…It describes frustration with the city…The city really, I don't think, did much to encourage the math center…If she …found someone who …had the money to rehab it…it would have been a wonderful thing… (Fig. 4.43).

Older adults felt that vacant businesses not only cast a bad light on the image of the community, but added to the perception of advanced decay and criminal activity.

…there aren't lots of vacant buildings… glass is broken in it…boarded up…I would hope that someone would want to open a business …Sad that it sat empty for some time…No progress seems to happen…They are easier to break in…Kids see that as fair game. No one's going to be…around there to even care if it does …If that's all you have to go on, you could think that may be an indicator of crime…I don't know what has kept anybody from moving in…I don't know if there are landlords that don't really care or rent is too high… people are always looking to open businesses and that one is sitting…vacant for at least a year… (Fig. 4.44).

This is the Chandler Construction Company Commercial Building …It has been in disrepair for …years. The roof is caving…Someone has ripped siding off to expose a window on the side…It's extremely dangerous looking…It looks like that street is in decline. The rest of the property owners keep their homes up…It's a terrible reflection in the city. It's depressing. It looks like it does not belong in the City of Ferguson… (Fig. 4.45).

Fig. 4.43 Math Center

Fig. 4.44 Vacant commercial building

Fig. 4.45 Chandler construction building

It is not only vacant homes and businesses that are problematic, vacant land can also become a symbol of community decay if it becomes an area for illegal dumping and other criminal activity.

> The airport came through…bought all of the buildings across the street…That's why it's vacant. They took up all of the streets…sidewalks…after the airport bought it, it was looking like a park…Then North Park Development bought it…took out all the trees…now it looks like the desert…There was actually a big apartment building right around the corner; an eight-family building. Across the street there were like three three-family duplexes… lots of activity…It's not too bad if they maintain it…this year…it's been okay. They cut it once, which is a milestone…They don't pay attention, act like it isn't there unless you call…They will send someone out…chop the weeds down …leave them in the street, and go away; cut the paper up…let it fly around (Fig. 4.46).

Fig. 4.46 Vacant lot

Racial Integration and Community Transformation

Many of the older adults who participated in the research study were lifetime or longtime residents of Ferguson. They remember when the community was predominantly White and have witnessed the racial transformation that occurred. However, most felt that the change was not something they considered disruptive. Older adults described their community as "integrating with variety," and the transformation as "positive." One participant described the individuals moving into Ferguson as "icky" and contributing to "increased crime."

> ...the diversity in Ferguson has changed amazingly...usually not in a bad way. When I was growing up...there were no Black people in Ferguson...now I think we are about 60/40... over a 50 year period.

A White male felt that his community continued to become more and more integrated throughout the years with a variety of races emerging.

> ...the neighborhood has become more integrated...both races moving in and out...That's a positive...The rest of the city is still becoming more integrated...it's not just one race coming in...It's still integrating with variety. Which is good.

Two White research participants, one male and one female moved to Ferguson from another community because they valued the diversity that Ferguson had to offer and wanted their children to have a broader education.

> ...we moved here at a time when the Ferguson and Florissant school district was going through voluntary desegregation...Everyone thought we were crazy...You know, you're going to have problems...We wanted a community that was not lily white. We wanted a community that was diverse...we have continued to value that.

One Black female spoke about the importance of churches integrating and coming together for various events.

> ...We are integrated...I love that...There will maybe be 20 churches that come together. We call it a church exchange...There's no Black or White congregation ...Everybody comes together.

This research participant also spoke about the importance of integration in her neighborhood.

> My neighborhood is not all Black ...very much integrated...so I like it.

However, not everyone interviewed felt that racial integration was a positive change in the community. One White female perceived that the racial transformation of the community brought some socially undesirable individuals into the neighborhoods.

> Trash is all over the place...People just throw it down...It's mostly the Black people I see doing this...It's a cultural these people have come from airport buyout in Berkley and Kinloch. Very much I mean always consider those icky...people I think more recently have come from the city...They tore down Pruitt Igoe...A lot of these people never have bought a house...They don't know how to take care of it. They didn't know they have to pay their rent or their house payment...This has always been a target because we have a lot more reasonably priced houses.

> All of the changes over the years (particularly our demographics) have contributed to increased crime.

Two White females also noted that it was difficult to attract desirable businesses to the area because of the change in demographics. They felt that White residents moved to more affluent communities with better schools.

> ...when I grew up here...this was an all-White community...White people are comfortable living...with a 28 % Black minority...We are now at over 70 %...we're in the minority... businesses look up their demographics...and they aren't coming here.

> ...Our public schools are populated mostly by African American children. I know many former public school teachers who have given up because of behavior and learning problems in the schools...Young professional and middle class couples do not wish to purchase homes here...They must send their children to private or parochial schools... The entire school issue is a major cause of concern...

However, one White male felt that although white flight had become the norm in certain areas of St. Louis County, he did not feel it applied to Ferguson.

> The trend in St. Louis County, for entering communities, is white-flight. The vacuum is almost deafening...the infrastructure, crime, so forth. It deteriorates very quickly. That has not been the case in Ferguson. And you see that by God's grace or careful planning of both of our leaders, there seems to be a nonaggressive but determined feeling from people that care and have a perception of place...thus so far are trying to make the effort to make it better.

Natural Surveillance

Male and female, Black and White older adults were apprehensive about obstructions that hindered their natural ability to observe activities within their neighborhood. This might include lack of street and porch lighting and unbridled vegetation growth that conceal nefarious activities from public view. Homes without lighting were described as “dark” and “uninviting.” Other overgrown areas “hide activities” and “intimidate” older adults.

One research participant described a vacant home that had no lighting because of the lack of electricity present in the house.

> …porch lights on the left…porch lights on the right…nothing in the middle…Well, houses that don’t have porch lights on…it becomes dark, it’s inviting…When there’s no porch lights, vacant houses quite often don’t have electricity on and so criminals see those Black houses and they become targets…Burglars that’s don’t want to worry about having to meet people while they are in the house…Older adults were also concerned about lighting in some of the neighborhood parks. Parks had specific hours to prevent individuals from entering them in the evening hours, however they admitted that people went there at night and participated in activities that hid them from view.

> It’s a pocket park that looks ok in the day time. There’s no lighting at night…a nice dark area to conduct elicit business. The addition of some solar powered LED lights might make a difference in being able to see…There’s some opportunities that…could be taken to improve its… function and its visibility… Neighbors across the street would be able to see more of what’s going on… (Fig. 4.47).

Research participants were also concerned with the condition of some lots and land in their neighborhoods. In particular, it appeared that mowing an abandoned vacant lot was part of the city maintenance code, but trimming of landscaping was not. Older adults felt that this should also be included in the code so that criminals would not have places to hide.

> …a deep empty lot that has overgrowth on both sides which effectively hide activities from being viewed from the street… The grass still gets mowed, but there’s isn’t trimming of overgrowth done here. That’s not part of the part of the code…if I had an empty lot next to my house I would be looking at what would it cost me to obtain that lot to make it part of my property so that I could eliminate that issue…Cleaning up the over growth on both sides so that a clear visual sideline could be seen from the street (Fig. 4.48).

Another common concern was overgrown vegetation around homes that concealed individuals from view who might have crime in mind. This included both property crimes and crimes toward people in the community.

Fig. 4.47 Pocket park

Fig. 4.48 Overgrown lot

> …the person can be standing behind the tree and I wouldn't be able to detect it until I was close up…I'm very cautious about my surroundings…I would trim back the tree and bring it up where it would not obstruct my view…During the day they can see me coming and be on one side and then as I'm passing by…they can come around and I would never see anything… There have been times when somebody kicked in a person's front door…that would prevent a neighbor from actually seeing it. We'd hear the noise but we won't see nothing… (Fig. 4.49).
>
> …these bushes.
>
> If you're walking down the street they are unable to see what's on the other side. It intimidates me to the point that I would probably cross the street. Especially at night time… Years ago I probably would have walked down there and not give it any thought…but nowadays you looking at everybody and everything… I want to know what that code enforcers doing for real…We've had issues…it would attract… home invasion, burglary, care tampering because it's blocking the vision of the automobiles there…Assault because you can't see whose up on you until they're there (Fig. 4.50).

One research participant knew that overgrown landscaping was a safety issue in the community and local police officers had stressed the importance of properly

Fig. 4.49 Overgrown bushes

Fig. 4.50 Overgrown landscaping

maintaining it to not giving people places to hide around the home. However, she thought that in some circumstances, the overgrowth looked pleasant.

> Ok, way overgrown…The police always give out these helpful hints about how to make your house look lived in is keep your bushes cut back…If you're trying to talk about somebody who has nefarious activities could hide… Although personally I think it looks kinda nice…Just because it's over grown doesn't mean that I don't like the way it looks…It makes me feel like some strange old lady lives here and I think I'd like her!…I personally hate to think that you can't do this anymore…In the 70's this would really been ok… It's kind of wild, it's not neglected…Overgrown with a purpose… (Fig. 4.51).

Maintenance

Male and female, White and Black research participants were equally concerned about maintenance in their community. When older adults addressed the topic of

Fig. 4.51 Overgrown with a purpose

maintenance in their community, they used words and phrases such as "contagious," "ashamed," "your house should be an example," and "responsibility." Several older adults felt that how someone cared for their property was an indication of personal character and how much they care about their community. Deferred maintenance did not "send a good picture of the neighborhood" and sent the message that "they don't care about the community." They felt that homeowners had a responsibility to the community, "being a homeowner does not mean that you just habitat the home...If you choose to live in a community...you have responsibilities to other people..." Research participants were concerned that a lack of maintenance in the neighborhood would "impact my home values...the general neighborhood...the impression of the community." In contrast, research participants felt that when someone did take good care of their property it indicated that, "somebody has values...somebody cares...about the neighborhood...the community as a whole..."

Many research participants seemed to agree that they did not feel that "...there weren't a lot of things...inducive to crime in the neighborhood as much as individual homeowners not maybe taking the best care of the property." Research participants agreed that outdoor maintenance was contagious in their community. When homeowners in their community did not properly maintain the outside of their homes and did not follow exterior city maintenance ordinances, it also encouraged others to do the same.

> This is the house with all the trucks...two recreational vehicles... two cars...pickups... part of a man's body underneath one of the RVs repairing it... You cannot repair vehicles in your driveway...The city inspectors have not done their job...It makes me furious. I can't understand anyone needing or wanting this many vehicles...It draws the neighborhood down...Outdoor maintenance is always contagious...There is someone on your block that you are trying to keep up with...you keep repainting...planting new things, making sure your grass is cut...When you see slummy looking properties...why should I bother? ...The homeowner must think that his home is his castle...he can do what he chooses...It would attract...more hoosier trash (Fig. 4.52).

Research participants also admitted that sometimes, outdoor maintenance had little to do with violating the city's outdoor maintenance ordinance. It was simply a

Fig. 4.52 Home with numerous vehicles

Fig. 4.53 Bucket garden

matter of personal preference for the homeowner that others in the neighborhood might consider unaesthetically pleasing.

> …the front porch is loaded with junk…This bucket garden that she has in her front yard… the city has tried for years to talk to her about the condition of her property… There's no ordinance against having a bucket garden…Her husband died a couple of years ago and she promised she was going to move but she hasn't…You have to divert your eyes (Fig. 4.53).

In other cases, it seemed as though the condition of one's property may be due to differing ethnic and cultural values. Unfortunately, some saw the condition of the property as offensive to neighborhood.

> …all the junk on the front porch…She's an immigrant …She's a character…Her place looked so bad that a group of kids as a missionary project came from North Dakota or South Dakota… volunteering for some church here locally in Ferguson and cleaned up her yard…She considered that a point of pride…Why aren't you ashamed of that (Fig. 4.54)?

Older adults were concerned about other homeowners not being aware of the way that their home was impacting everyone else in the neighborhood.

> The blue tarp… somebody that is just not vigilant about caring for the property…Whenever you have a situation like this…it's not beneficial for the neighborhood. I don't think that some homeowners understand that they are not living in a vacuum. This is going to impact my home values…It impacts the general neighborhood …The impression of the community… (Fig. 4.55).

Fig. 4.54 Home with overgrowth

Fig. 4.55 Home with tarp on roof

...this is our neighborhood hoarder...He's got a vine that is...swallowing his house... Everything's just horrible...The city has... gone up there with crews... cleaned his stuff... taken everything out of his yard ...dumped it two or three times...He also owns another house...He would take everything out of that one...move it down to that one...They would inspect that...say 'ok well that's alright'...then he'd fill em' both up...It really does impact that lower block...Who would move next to that?...It frustrates me...the neighbors...the city...they don't know what to do with him. He's a nice guy if you talk to him...he could be on that hoarders show (Fig. 4.56).

Older adults were troubled by the enforcement of ordinances in their community. It was felt that the City of Ferguson had some very good ordinances, but they were either not being properly enforced or that homeowners were not taking the fines seriously enough.

...that scaffolding has been sitting there. For years...it is the next step to the house deteriorating those people don't care about how it looks...They don't care about the community...It's not safe or somebody could break in...We have some pretty good ordinances... They try to enforce them...If you go to a meeting where people are paying fines...they are packed... (Fig. 4.57).

...they certainly are not cleaning up their storm damage...The city...after 5 months ...still not forced them...to deal with it ...They cite them...The ordinance is very specific that you

Fig. 4.56 Neighborhood hoarder

Fig. 4.57 Scaffolding

cannot have dead… trees or fallen branches…unbridled growth …Your grass can't be taller than 7 inches…You'd think 'wow good ordinance'…we gotta enforce them. After a lengthy time the city will come in and clean it up…bill the property…It's always a combination of sadness and disgust. And when it gets bad enough outrage!…I am convinced that when cities do not have strict enforcement…there's a grapevine that goes out; people communicate…they say 'eh you're lax on maintenance, no problem. They'll never bug you…just be a slob and feel right at home (Fig. 4.58).'

…so overgrown I can't really tell whether anybody lives here…It's hard to even see the house…It doesn't look overgrown with a purpose…Don't care or else maybe they're incapacitated and can't care….What about the Ferguson rules?…for all these rules…they have they're either not enforced or…It just doesn't send a good picture of the neighborhood period…I gather that …the City of Ferguson does send people like this notices but if the fine is not great enough that they go the hell with it and don't bother. It's too much trouble to clean it up (Fig. 4.59).

One research participant was particularly concerned about the lack of and/or confusion about responsibility taken on some properties in or near the City of Ferguson. These properties were impacting homeowners because they were taken advantage of for illegal dumping.

Fig. 4.58 Storm damage

Fig. 4.59 Home with extreme overgrowth

> ...the overgrowth hangs over...You don't really have access to the full street...There's paper and trash intermingled...Looks disgusting...tires and bumpers...mixed in there ... Somebody needs to go...and clean it up. Nobody wants to take any action on it...Ferguson says it's Berkeley...it's not Berkeley...It's Cool Valley...It's a little irritating...It's not a major task if they clean that up...make the street look a whole lot better...I would clean it up...have somebody be responsible for it. Some of the residents have taken responsibility...and cleaned it up a little bit... The others just sit back and wait (Fig. 4.60).

Not all maintenance issues in the community were related to individual properties, but concerned aging infrastructure in the community.

> ...diminishing tax base, and the infrastructure suffering for it...For the portion of the community that the city is supposed to be taking care of, there's is just not enough money to support that...Some days...it doesn't bother me a great deal...other days, this place is going downhill...If you allow it to, it can make you hopeless...I've got some neighbors that think the city doesn't care...just looking for somebody to blame.

One White female believed it was the duty of homeowners, especially this well-known predominant city leader to set an example for others in the community.

Fig. 4.60 Overgrowth hanging into street

…that house is a real problem…The people that lived in this house, one of the reasons they moved is because of this homeowner. This house is just so annoying…I have spoken to this homeowner. I said to him…'your house should be an example'…you're not supposed to have your garbage cans in front of the house, there is a tree kind of growing out of the gutter…He has a broken window up here…It is reprehensible…The back yard at one time was overgrown…It is disgraceful…The things that would correct this don't cost a lot of money…I don't think it necessarily represents the neighborhood…The rest of the neighbors are very conscientious…It represents…a lackadaisical homeowner who knows better… Being a homeowner does not mean that you just habitat the home…If you choose to live in a community…you have responsibilities to other people and to the community as a whole…That's why it's called community.

Rental Property

Many research participants were disturbed by the number of rental properties in their community, the transient population associated with rental property, and how the lack of upkeep to these properties ultimately affected property values in the neighborhood. Research participants described renters as a "transient population," "don't care," "don't have the pride," and who "bring trouble." Participants were concerned with landlords because it appeared that some of them were not local landlords and were making money but do not "reinvest it back into the property" to maintain the home. If landlords were local, it was less of a problem. However, when landlords lived further away, there was an "out of sight, out of mind" attitude and it became much more difficult to track them down and hold them responsible. Research participants agreed that the City of Ferguson was beginning to take some very proactive steps to control the cycle of renters and to make sure rental property was properly maintained. One Black male stated:

…it's changed quite a bit in the last 30 or so years…It's gone from middle class to a major portion of rental property…

Research participants felt that because renters had no real investment in the properties, there was no incentive for them to take good care of them.

Renters. They don't really take care of their property…They don't have the pride to do anything…They really don't own the house…They just don't care…If you've got money in something, you're going to want to take care of it.

I think ownership is better than rental when you've got someone that owns. Typically, they would take better care of their house, than renters. Renters tear them up.

Research participants were also concerned about the transient nature of renters in their neighborhood. Often when the lease is up on the property; renters are moving on. This creates an unstable population in the community who does not have a solid investment in the betterment of the neighborhood.

Fig. 4.61 Yellow rental home

> ...This little yellow house ...is a rental house...that has been a problem off and on...Rental property is an issue...It's just this transient population ...Not a stable situation...It's very cute inside, but it's teensy ...We're probably not going to get a nice young couple with children. The original owners had kids, but...people had smaller houses then... (Fig. 4.61).

> ...our properties are rentals... transient I don't care...not investing myself in your community...inhabitants...I don't think that does the community any service...The percentage keeps going up...

Unfortunately, rental properties were often left abandoned because the amount of money needed to bring the property up to the standard maintenance code cost more than it was worth to the landlord.

> ...it's gotten progressively worse...You're seeing more...vacant houses...That's a landlord situation...In my neighborhood there's quite a lot of...rental houses...They're just sitting. The property values going down...What they paid for them is a lot more. They don't want to put any more money into them.

> ...the house is vacant ...The door it's in perfectly fine shape, there's no glass damaged... The only reason that the plywood is on there is to prevent something from happening. It makes me feel apprehensive...The owner of the vacant house took into his hands to put this plywood up...that drives the property value down of the occupied houses...He owned the house...was renting it, but he could no longer rent it to new people...He couldn't get his... landlord license renewed...You have to pass the exterior and interior inspections...It cost too much for him to do that (Fig. 4.62).

In addition to properties being left completely abandoned, some landlords were not quick to make improvements. One Black male felt that the landlords were not taking the money that they were making off of the properties and reinvesting it in maintenance.

> It's only that one particular property right there that was a...so to speak ghetto landlord. Rents the property out to make the money...then doesn't take the money that he makes off the property and reinvest it back into the property...Well, it makes me angry...I really keep a watch on the property... Whenever I hear a noise I...go out and check or I'll call the police if it's really loud.

Fig. 4.62 Boarded up; abandoned rental

Another common theme among abandoned rental properties were the neighboring community members who looked after them to make sure that no nefarious activities occurred and that some minimal maintenance transpired. Primarily males, both White and Black, took on this responsibility.

> …it is an empty house….a rental property…The landlord has been very slow about making any improvements to it…The guy that lived there before let a few things go down…didn't keep it up…We all kind of take a turn around the house to see if there are any open, broke windows, any evidence like…anyone's broken into it…It's kind of odd to have that piece of the puzzle…not filled in…See probably some cars in the driveway…kids playing (Fig. 4.63).

> …the house next door…He was renting it out…since she's moved out nobody's moved in…The house is up for either auction or…quick sale…Being that's there no one here to take care of this property I do what little I can to keep it clean…to keep…my property from looking so bad.

One Black male was also concerned with the cultural values that renters brought to the community.

> a lot of times it appears to me that the neighborhoods up for sale…Being with present economic condition the banks are now talking about renting homes out instead of selling

Fig. 4.63 Vacant rental home

> the homes…If you move somebody into the home that doesn't have the same values as the residents that are living there…it messes up the neighborhood. It messes up…the community…You've got two different sets of people…people who don't have values equivalent to the people living there who want to take care of their property…Want to see the neighborhood look better. Keep it clean and livable.

A Black female felt that certain price points on property brought in types of people with dissimilar values.

> …certain prices…bring certain type of people. If it's Section-8, you bringing other type of people…bring the wrong group of friends to the neighborhood…bring trouble when you're a homeowner…I got away from the city because of things like that so I don't want to turn around and have to face the same thing…I'm invested in my property.

However, one White male felt that the City of Ferguson was taking very proactive steps to control the cycle of renters in their community and encouraging them to become better renters.

> The city is…taking some very proactive steps in rental housing, to control the cycle of people moving in as renters that maybe haven't had a background in home ownership or maybe their extended family is all the culture that they know. Sometimes that brings in people…that bring bad things in with them…That was very proactive and very wise of our city leaders.

In addition to helping control the cycle of renters in their community, one Black female also felt that the city was taking various proactive measures to try to hold landlords more accountable for the properties they own.

> When we rezoned…we were going to hold landlords accountable. Commercial landlords… You cannot tell that those businesses are not doing business, except for the sign saying for sale or lease. The area does not look abandoned…They don't intend for it to become an issue…I can tell by how much they monitor me…They are doing the same to everybody else.

Another Black female also felt that some of the petty crime that occurred within their neighborhoods was due to the transient population of renters not invested in their community.

> …that rental property…just loud…the people that they bring to visit. Too many cars all in front of the neighbors' homes…When you see a lot of vacancies…they are out of rental… After the year lease is up they're moving out…another families moving in. We have periods…that is somewhat disturbing…it's due to rental property…having little break in and stuff.

Synthesis of the Experience

The true essence of the phenomenon is maintenance or "to maintain." The Latin word for maintain is manu tenere, meaning to "hold in the hand (Merriam-Webster, 2014)." As people age, their emotional grasp tightens, holding onto that which is still in their control at a time when control is slowly slipping away. They want to "hold in their hand" the nostalgic image that was Ferguson when they were growing up and preserve it as they continue to grow older. One research participant so eloquently captured this concept through his spoken words.

> I need to say this about the audience that you have been seeking out. When your years get less, you've got more behind you than you've got in front of you. We all start feeling our mortality a little bit and that's frustrating…When we were younger we thought we were more tolerant…now we are finding out that we're not. We like to have more control over something. So a lot of the belly aching you hear, it seems related to the fact that we have less and less control over our lives. Well, the perception is that control is slipping away from us. And this is belly aching about something like this is a good relief.

It is quite simplistic to equate *maintenance* to the way that a particular piece of property is cared for and how this affects the perception of crime in the community. However, *maintenance* of physical, spiritual, mental, and social well-being becomes increasingly important as older adults work to combat some of the fears and frustrations that they experience from the perception of crime in their community. Connections between friends, family, and neighbors must be *maintained* to not only determine and preserve the social norms of the community, but also to establish natural surveillance within the neighborhood to protect individual investments. As demonstrated, many things in Ferguson have not been *maintained*, but have instead dramatically transformed. Racial homogeneity, the integrity of the housing stock, and many of the fundamental small town businesses that attracted and retained older adults in the community have not been *maintained*. However, older adults' attachment to place has been *maintained* stronger than ever, in addition to their *maintained* physical and social connection in the community. The preservation of many historic features throughout the community have been *maintained*, as well as the picturesque natural landscapes, giving older adults throughout the community an opportunity for mental restoration that prepares them to deal with the daily challenges of community life.

For these older adults, their subjective perceptions illuminate the objective reality captured through their personal photographic illustrations of the community. Their insider perspectives demonstrate virtually unproblematic personal safety concerns. Nevertheless, the contagious nature of poor private property upkeep exemplified through inadequately cared for and dilapidated rental houses, vacant properties, single-family dwellings, and locally owned businesses; symbolizes a cultural insensitivity to the consequences of damaging outsider perceptions regarding the safety of their community. Older adults' personal concern with the generational deterioration of values appears to manifest itself through undesirable actions and encounters in the community. Particularly, littering and illegal dumping became a reprehensible exploitation by both community insiders and outsiders; symbolizing an objective devalued perception of their dwelling place as well as a subjective diminishment of the spirit to those who call Ferguson home.

Racially integrated community churches, businesses, and schools presented a perceived positive experience for many; providing opportunities for diversified involvement. Conversely, it appeared to others a historical trauma to the once all-White community; adversely affecting public safety, contributing to a failing school district, undesirable personal property care, and an inability to attract perceived reputable businesses to the area. The eyes and ears of their neighborhoods, elders detect and report suspicious activities, keeping a keenly watchful eye on neighboring properties. Many, with a long-standing physical and emotional bond become instinctively in-tune to the slightest subtlety; thus enhancing the natural surveillance of the community. Apprehension becomes additionally apparent as they observe obstacles within and around their neighborhood, effectively concealing nefarious activities and altering their proficiency to naturally survey the neighborhood. The union of picturesque urban parks, trails, gardens, and markets with personal narratives thread past youthful remembrances to a present day utopia, and onto progressive-futuristic opportunities to further unify the community with nature. Inspiring accounts of day-to-day personal and entrepreneurial investment signifies a community laboring to revolutionize an identified tarnished image to one that inspires individuals to grow up and grow old within their neighborhoods. The presence of historic community elements become a fundamental foundational root in Ferguson, conjuring up nostalgic images for those growing old in the community. When well maintained, they become a beacon of hope, representing a long-established element unchanged by the tides that are the chronology of community transformations.

Older adults have verified the significance of community perceptions through the photographs of their world. As insiders in the community, these older adults comprehend the stable firm foundation that is Ferguson. Nevertheless, they recognize the power of perception to those viewing Ferguson from the outside. They recognize the past decades of transformation that have shaped and molded Ferguson into what it currently is. They (community members, the City of Ferguson, and outside investors) recognize the image that Ferguson is working to improve. They do not ignore this image, they recognize it, and work to build a stronger Ferguson. As demonstrated, the presence or absence of actual community safety can be concealed or exemplified through images of deteriorating infrastructure, symbolizing a larger social decline. Conversely, carefully maintained homes, streets, and business districts symbolize that someone cares about the image of the community, and thus must care about the safety of the people who reside and visit.

References

City of Ferguson. (2015). *Ferguson neighborhoods.* Retrieved from http://www.fergusoncity.com/documentcenter/view/1551property

Ferguson, MO. (2015). *Google Maps.* Google. Retrieved from https://www.google.com/maps/place/Ferguson,+MO/@38.7489992,-90.3241687,13z/data=!3m1!4b1!4m5!3m4!1s0x87df49e36de09f05:0x7b87d2fc9ee09289!8m2!3d38.7442175!4d-90.3053915

Chapter 5
Discussion

The purpose of this research study was to learn more about older adults' perceptions of crime in Ferguson, Missouri and how their quality of life was affected. The results of this research study indicate that concerns with crime in the form of neighborhood disorder do exist. However, the majority of older adults did not feel that their own personal safety was compromised and did not worry about personal victimization. The presence of lower level neighborhood disorder created some anxiety about whether signs of physical and social deterioration suggest the onset of future significant decay in the community. Race seemed to play a contributing role in how older adults interpreted lower level disorder in their community. However, socioeconomic class was potentially more significant in how they interpreted physical and social disorder. These older adults' attachment to place combined with their social and physical investment in the community seems to counter the fear of crime they could experience from the presence of neighborhood disorder. Thus, their quality of life has not been gravely affected.

The Social Disorganization Theory and Broken Window Theory provide the main theoretical perspectives supporting the development of the results in this research study. Additional theories (Defensible Space Theory, Environmental Docility Theory, and Place Attachment) play supporting roles in explaining some of the intricacies identified through the individual thematic areas. The textural and structural descriptions within the 12 thematic areas interconnect with the theoretical philosophies to explain the dynamic community relationship and provide a platform for discussion and interpretation of the research results.

As described by Clare and Plecas (2012), disorders (social and physical incivilities) are "low-level breaches of community standards that signal an erosion of conventionally accepted norms and values (LaGrange et al. 1992, p. 312)." Clare and Plecas (2012) also suggest that obvious visual symbols of neglect such as decaying buildings, littering, graffiti, and vandalism symbolize physical disorder. Particular behaviors and attitudes, such as loitering, public drunkenness, and intimidation become signs of social disorder (LaGrange et al. 1992). Felonneau (2004) defines incivilities "as acts of nonrespect or of aggression towards others and

K.L. Fisher-McLean, *Perceptions of Community Crime in Ferguson, MO*,
SpringerBriefs in Criminology, DOI 10.1007/978-3-319-46786-3_5

towards the environment, without necessarily reflecting the legal category of delinquency or criminality (Furstenberg 1971; LaGrange 1984, 1995; Roche 1996, 1998). Such behavior are generally not very serious but become unacceptable on a daily basis (Wievorka 1999) because they trigger a diffuse feeling of the decline of the social order and there a sentiment of fear (Moser and Lidvan 1991; Derbarbieux 1996)." Ross and Mirowsky (1999, p. 413) suggest that, "physical disorder–such as litter, graffiti, and vandalism–also indicates that social control has broken down" (cited by Austin and Sanders 2007). Austin and Sanders (2007) additionally suggest that "perceived disorder is also associated with erosion of social ties (Ross and Jang 2000), which may in turn impact satisfaction with the community (Ladewig and McCann 1980; Wasserman 1982)." When informal connections are maintained between one another, community members can combat the effects of neighborhood disorder, "reducing fear and mistrust," and those who are more trusting and not fearful of one another, bond even more closely, establishing greater "social control in the community" (Austin and Sanders 2007).

Nearly all of the elements older adults in Ferguson spoke about that contributed to the perception of crime in their community can be considered lower levels of physical and social disorder in their community. These include properties that are vacant and/or rundown, littering, loitering, and other various elements that signify that Ferguson may not be as preferred as a place to live or visit as it once was. This perception is also potentially exemplified by the belief that their environment, as young people growing up in the community, was more socially stable and physically secure than it is now (Dionne 1998, p. 93). However, older adults revealed as many positive and promising aspects about their community. These include opportunities for physical and social engagement, opportunities for access to nature, the effort to preserve historic sites throughout their community, and new investors attracted to Ferguson to shape the future commerce and vitality of their community. The positive opportunities offered, help to counter the effects of the physical and social disorder that exist.

Physical and social engagement in the community can formally and informally occur. Formal opportunities for physical and social engagement include attendance at neighborhood watch meetings, city council meetings, gardening clubs, the farmers market, historical preservation groups, and other such organized activities, where they get opportunities to meet members of the community who share similar interests. Physical and social engagement opportunities identified by older adults that contributed to informal social connections in their community included walking, biking, and jogging through the neighborhood and on trails that offer them opportunities to recognize faces, activities, behaviors, and informally meet neighbors, their children, and pets in a less prescriptive manner. Both formal and informal social connections provide older adults with the opportunity to discuss physical and social disorder in their community. This is not only how they generate their own social norms about their neighborhood, but also how they learn how others view the social norms of their community. They then decide what might constitute social disorder in their community and possibly what can be done about behaviors and actions that do not fit the social norm. They can also collectively identify signs of physical disorder in their community, what might be done about it, or what it might signify. Thus, these collective actions begin to establish and maintain social control in the neighborhood and community.

Social Disorganization

The above interpretation of the phenomenon ties directly into the Social Disorganization Theory. Serious crimes are often the focus of Social Disorganization research rather than more minor offenses. However, Markowitz et al. (2001) acknowledge that minor offenses often contribute heavily to the fear of crime that neighborhoods experience (cited by Porter et al. 2012). On one end of the spectrum, communities are socially organized; on the other end of the spectrum communities can be characterized as social disorganized. Characteristics of a socially organized community include an agreement about norms and values, social integration amongst residents, and solid connections between residents who live in close proximity to one another. The support of these characteristics often leads to lower crime rates in the community (Kubrin 2009). Kubrin (2009, p. 227) also describes that in organized communities, "there is evidence of (1) informal surveillance, or the casual but active observation of neighborhood streets that is engaged in by individuals during daily activities, (2) movement-governing rules, or the avoidance of areas in or near neighborhoods viewed as unsafe, and (3) direct intervention, or the questioning of strangers and residents of the neighborhood about suspicious activities, chastening adults and admonishing children for behavior that is defined as unacceptable (Greenberg et al. 1982)."

As cited by Sampson and Groves (1989), evidence of "low economic status, ethnic heterogeneity, residential mobility, family disruption" and a lack of "effective social controls" increase social disorganization in the community (Kornhauser 1978; Bursik 1984). For those who participated in the research study, several elements associated with social "disorganization" were demonstrated. While some of these physical and social incivilities affecting social organization were directly observable and easily communicated by older adults, some were subtler changes in the community. For examples, loitering, playing basketball in the street, youth walking in the street, and changes in attitude and demeanor may encompass a disagreement about what are acceptable community social norms, thus creating social disorganization. Acceptable social norms may be transforming as an older White generation grows old and a younger Black generation moves in. After the fatal shooting of Michael Brown and initiation of community turmoil, the nation began to see the scale of social control tipping as disorganization seized the moment. Those who felt powerless dominated the streets of Ferguson and other cities around the nation, questioning what they felt were the unacceptable social controls and standards put in place by those in power.

At the time of data collection for this research study (October 2013–June 2014), older adults demonstrated many examples of ways that social "organization" were upheld. These features associated with community organization help to maintain current social controls and standards and curtail aspects of perceived social disorganization from dominating the community. One such way that older adults demonstrated social organization in their community was by directly questioning strangers and setting an example of appropriate norms and behaviors for the neighborhood. A Black female described a man in her neighborhood seeking out directions:

> There was a new guy that I never seen when I was out on this day…He was trying to get to Georgia…I said…'you have to go back to Harvey…make a right…then you will be on Georgia'…It never crossed his mind to approach me…What…I did…was let him know… in this neighborhood we are approachable….Guess what, the old lady knows how to get you where you want to go.

She may have viewed her actions as cordial. However, what she unintentionally accomplished was an exhibition of social control by sending the message to the stranger that people are informally watching out for their neighborhood and can identify those who do not belong. Because she conversed directly with him, she would also be able to get a feel for what his possible intentions within the neighborhood were and be able to easily identify him in the event that it became warranted.

This same research participant also demonstrated social control through her interaction with young adults, helping to establish and preserve the social norms and values of the neighborhood.

> The little guy on the corner, he's been there for about nine months…I heard this 'baboom, baboom, noise'…I went over in my robe and…said, 'hey, this is a neighborhood. You can't have this noise here…I don't want to have to put a sign up about noise.' 'Oh, no ma'am, no ma'am.' I said 'okay, tell all your friends'…He told all of his friends don't come here with all that noise…They work in a band…They are real good, too…But you're not going to play over here like that.

However, she also demonstrated that when these social norms and values are respected among the neighborhood, solid connections are strengthened and mutual respect is established, reinforcing the social organization of the neighborhood.

> …when he first got his lawn mower, he came over…asked my husband to show him what to do…He bought this big barbeque grill…he called my husband over and said 'can you come over and show me how to use it?'…When he had to go somewhere where he had to wear a tie, he came over…and showed him how to tie the tie…

Another element described as contributing to social disorganization in the community was the presence of ethnic heterogeneity. Nearly every participant noted the dramatic shift in the racial demographics of their community. Most research participants felt the integration was positive, some did not.

> …when I grew up here…this was an all-White community…White people are comfortable living…with a 28 % Black minority…We are now at over 70 %…we're in the minority… businesses look up their demographics…and they aren't coming here.

For them, they felt that the social control of their community was shifting along with the social norms of the community they once knew. The social norms were beginning to resemble that of other deteriorating inner ring suburbs and inner city neighborhoods and not that of which attracted them to and kept them in Ferguson for the long haul.

Social Disorder

In a study conducted by Kelling and Wilson (1982), Kelling invested hours walking with a Newark, New Jersey police officer "to see how they defined 'order' and what they did to maintain it." One of the roles of the police officer was to not only keep order in the formal legal context, but to keep order in the sense of informal but also "widely understood" rules within the neighborhood or community. For example, "drunks and addicts could sit on the stoops, but could not lie down. People could drink on side streets, but not at the main intersection. Bottles had to be in paper bags. Talking to, bothering, or begging from people waiting at the bus stop was strictly forbidden." The definitions of the social rules were decided by the behavioral norms of the rest of the neighborhood. Much of what the police officer was doing could probably not be described as "enforcing the law," but involved steps to protect the informal rules established by the neighborhood itself. In this study, the people of Newark, New Jersey "assigned a high value to public order" and were alleviated when police maintained that order. This is because disorderly streets in the form of "distasteful, worrisome encounters" contribute to fear and anxiety about potential crime.

Undesirable Attitudes and Behaviors

Older adults were also concerned with other low levels of social disorder such as loitering, particularly by youth walking down the middle of the street rather than on the sidewalk. This was a characterization made by both White and Black female research participants.

> I personally am not comfortable when I see groups of African American youths walking up and down the middle of the street...They're always screaming at one another or fighting.

> The groups...intimidating to seniors...walking in the streets instead of on the sidewalk.

As part of the recruiting process, I attended many neighborhood watch/community meetings to try to engage older adults in the research study. Although I was unable to recruit any older adults from the predominately Black Nesbit Newton neighborhood, I listened to their many concerns intently. The Nesbit Newton neighborhood is also located adjacent from the Canfield Green Apartment

Complex. Residents of this neighborhood were also extremely concerned with the presence of loitering teens and young adults. I distinctly recall them asking the police officer present at the meeting why they no longer stop and question the groups of loitering teens. The residents stated that these groups often caused problems for the neighborhood and were involved with drug dealing and other illegal activities. The police officer responded to the group that his officers were no longer instructed to stop and question groups of teens without a reasonable cause because this was an action that could be seen as profiling (racial, gender, etc.). Despite what the officer had told them, these Black residents were encouraging law enforcement to take action against loitering teens in their neighborhood.

Two months after the conclusion of the data collection for this research study, Michael Brown and Dorian Johnson were initially stopped by police officer Darren Wilson for walking down the middle of the street near the Canfield Green Apartment Complex. Although not extremely criminally negligent, the behavior was identified through this research study as socially disruptive by older adults. Therefore, it is possible that Office Wilson's initial contact with Brown and Johnson related to maintaining the current overall informal *community* social standards even when the social standards in this particular *neighborhood* had already shifted.

Community Image

Research participants in Ferguson also seemed to assign a high value to public order in their community. Several of the social incivilities older adults were concerned with in their community could be described as the perceived violation of the social, informal, unspoken rules established in their community, such as loitering youth, the presence of homeless people in their community, the perceived infiltration of undesirable businesses, and the exodus of more desirable businesses from their community. One research participant described the appearance of homeless people in Ferguson as unattractive. She noted, that some appear to be mentally unstable, making people passing by uncomfortable. This added to the perception that Ferguson was unsafe. However, she acknowledged that she did not think that homeless people in the community were responsible for crime. A study conducted by Newburn and Rock (2004) investigated violence and victimization among homeless persons in the United Kingdom. They suggest that homeless people are often seen as crime committers, when in fact, they are far more likely to become victims of crime than to actually be an offender. A quantitative research study conducted by Gallup, Inc. on behalf of Mae (2007) also investigated America's perceptions and attitudes about homelessness in America. Among their findings, "one in every two adults surveyed (51 %) completely agree that 'communities are safer when people don't have to live on the streets' and fewer than half (44 %) completely agree that they 'want to live in a community that provides for the care of its homeless citizens.'" According to Donley (2008), people seem to associate criminal behavior with homelessness (Barak 2002), whether or not it is "empirically supported." Mostly people are fearful of what they view homeless people as: "dirty, smelling of alcohol or carting a multitude of belongings, they cannot easily blend in with the people around them…they may stay for hours on a bench, in a public library or in a park. Their visibility, appearance and demeanor make many people nervous" (Donley 2008).

Another concern among some of the older adults was the presence of what they considered marginal shops, such as beauty supply stores, liquor stores, and traffic law centers that catered to "poorer demographically and African American's" and added "an element that is not healthy for a community." Beauty supply stores throughout the St. Louis area have also been targeted for theft of hair extensions, which can be easily stolen and resold. "They're relatively small and easy to grab, and have no codes or serial numbers that can be tracked" (Bell 2013). Therefore, the concern amongst older adults may also tie into the fact that this type of store may attract some additional "actual" crime to the community. A couple of research participants were also concerned that their community could no longer attract quality businesses, such as department stores, hardware stores, and reputable dentists and doctors to the area because of the changing racial demographics of the community. However, research participants also acknowledged that they felt Ferguson was turning the corner. It had begun to attract and retain new restaurants to the downtown area.

Although both Black and White research participants felt that Ferguson did not have the quality businesses that it once had, it was only White females who acknowledged the shift in racial demographics as contributing to the decline. Interestingly, both a White and Black female research participant spoke about what appears to be the same liquor store in their community. However, they each frame the scene in a very different light. The first is an abbreviated account told by one of the White female research participants and the second account is told by one of the Black female research participants.

> …a liquor store…that disturbs people…They feel it attracts an element that is not healthy for a community. A group of friends and I used to go up every Sunday morning and pick up litter…We finally stopped…It was so hopeless…After about 3 or 4 years…we quit… discouraging.

> Most people think when you see a liquor store there's all the people hanging out…People would…shy away from there…I'm not shying away from there. I'm going up there. I'm passing there…I got a lot of respect from those people that stop in there…A little corner store for a lot of people that don't have cars…It's so orderly and clean that I don't have a problem passing it.

In a study conducted by Sampson and Raudenbush (2004), they found that "neighborhood racial context helps shape residents' perceptions of disorder…Black residents reported lower levels of disorder than White residents for the same neighborhood." This could help to account for differing perspectives about the presence of the liquor store in the community. The White older adult female potentially saw this as a violation of her perceived social values and norms, thus signifying disorder and compromising community safety. The Black older adult female described the liquor store from a positive perspective. She refers to it as a corner store and acknowledges that it serves as a viable shopping option for those who do not have a vehicle to get to a larger supermarket.

As cited by Small and McDermott (2006), "the de-institutionalization of neighborhoods is the product of middle-class flight and the ensuing concentration of poverty. Because middle-class residents economically sustain neighborhood establishments, their absence undermines the latter's stability and survival (Wilson 1987, 1996)." Pollard (1996) finds that in South Central Los Angeles, as the number of "poor and non-White" individuals grows, the quantity of banks decreases and utilizing figures from "metropolitan Louisiana and Cook County Illinois," Graces (2003) found that "'payday lenders' such as check cashing establishments are especially likely to be found in low income non-White neighborhoods" (cited by Small and McDermott 2006). There is also a disproportionate amount of grocery store in "poor communities of color, compared to wealthier, whiter communities" (Flournoy 2006). The Center for Food and Justice at the Urban and Environmental Policy Institute conducted a study and their findings indicated that "middle-and upper-income communities in Los Angeles County have twice as many supermarkets per capita as low-income communities...predominantly White communities have three times the supermarkets of predominantly Black communities, and nearly twice those of predominantly Latino communities" (cited by Flournoy 2006). However, according to Morland et al. (2002), "research on 'negative' businesses such as bars and liquor stores finds that poverty increase their prevalence" (cited by Small and McDermott 2006). Therefore, it could be hypothesized that the infringement of shops such as liquor stores, beauty supply shops, and other such businesses are viewed as attracting low-income minorities. This makes some older adults perceive the infiltration not so much as a threat to their personal safety, but as a threat to their community's social norms and the intrusion of a less than desirable population in their community. As previously noted, the acceptable social norms and behavioral standards of the community are changing as the demographics of the population change, becoming uncontrollable and unpredictable to some older adults.

Rental Property

Older adults were concerned with the number of single family homes that had become rental property in their community. Particularly, they were concerned with the lack of maintenance to rental property, the transient nature and differing values of renters, the type of people (friends and family) that renters brought into the neighborhood, the nature of crimes associated with the increase of renters throughout the neighborhoods, and the accountability of absentee landlords. This also ties into the Social Disorganization Theory in which residential mobility leads to social disorganization in the community. Although various rental properties were mentioned throughout the City of Ferguson, two older adult White males mentioned the apartment areas off West Florissant Avenue and the problematic issues associated with them.

> There are areas that are a part of the City of Ferguson that are probably more problematic… apartment areas…across West Florissant that used to be called Canfield Green…There have been lot of troubles in the past, but recently hasn't…There's the Sharon Dale Apartments behind McDonalds over by West Florissant, but they're putting a lot of money into. They put up wrought iron, six or eight foot tall fences all around the property…It's nice looking…I would feel safer…The fences would prevent someone from parking along… Ferguson Avenue…running into the apartment building…I see that as some real improvement to those apartment areas.

> By the transient nature of the apartment buildings that we have…There's robbery… shootings…domestic violence…It's unfortunate…It's primarily on the victims…who live there…It's self-predatory…An unfortunate situation…The city can send our cops in…I think the most frustrating thing is…it must be terrible to be in a mindset that is so hopeless that you allow that to continue…It seems like…people involved with that would say enough of this. But it doesn't occur…there must be powerlessness that goes along with that. I don't know how to image that.

Unfortunately, these apartment complexes were the location where the deadly encounter took place between Officer Darren Wilson and Michael Brown. The incident escalated when Brown attacked Officer Wilson and as a result, Brown was shot and killed in August, 2014. Older adults believed this was an area of Ferguson where more crime existed. As reported in 2010, Ferguson law enforcement was requested at apartment complexes such as Canfield Green nearly 50 % of the time (Bogan 2014). In the early 1970s, the United States Government established home programs such as Section-8 following the demolition of Pruitt Igoe in North St. Louis. Section-8 vouchers, which have doubled since the mid-1990s according to the Department of Housing and Urban Development, provide renters with the option to live within the community rather than in large scale apartment complexes. Concentrated in limited income communities, many inner North St. Louis County suburbs such as Ferguson appeal to Section-8 tenants (Bogan 2014). Pruitt Igoe was located approximately 11 miles from Ferguson, Missouri. Acknowledging these changes, one White female felt that the Lambert Airport expansion as well as the destruction of public housing projects such as Pruitt Igoe displaced residents into the nearby suburban communities.

> Trash is all over the place…People just throw it down…It's mostly the Black people I see doing this…it's a cultural these people have come from airport buyout in Berkley and Kinloch. Very much I mean always consider those icky…People I think more recently have come from the city…They tore down Pruitt Igoe…A lot of these people never have bought a house…They don't know how to take care of it. They didn't know they have to pay their rent or their house payment…This has always been a target because we have a lot more reasonably priced houses.

Although Ferguson is one of the most racial diverse suburbs in the area today, that has not always been the case. According to Rothstein (2014), the primary thoroughfare between Ferguson and Kinloch was chained off to bar most African Americans from entering Ferguson up until the mid-1960s. Many of the older adults who participated in the research study embrace Ferguson's diversity and identified racial integration as positive for their community. However, some did not and felt changing demographics contributed to increased crime, littering, and the exodus of desirable businesses from the community.

An older adult Black female who was not a lifetime resident of Ferguson also stated her concerns with Section-8 renters in her neighborhood. She knew of a home near her in which they were no longer accepting Section-8 renters because of the problems they had caused. Renters, in general, many of whom are low income, were perceived as bringing the "wrong" values to the community. They were perceived as not being invested in the future of the neighborhood, and presented various problems for homeowners.

> ...certain prices...bring certain type of people. If its Section-8 you bringing other type of people...bring the wrong group of friends to the neighborhood...bring trouble when you're a homeowner...I got away from the city because of things like that so I don't want to turn around and have to face the same thing...I'm invested in my property.

An older adult Black male stated:

> ...if you move somebody into the home that doesn't have the same values as the residents that are living there...it messes up the neighborhood. It messes up...the community... You've got 2 different sets of people...People who don't have values equivalent to the people living there who want to take care of their property...Want to see the neighborhood look better. Keep it clean and livable.

Both of the above older adults were not lifetime residents of Ferguson, Missouri. At one time, they had both resided in the City of St. Louis. They left the city and it is possible that they chose Ferguson because in comparison to other North County St. Louis suburbs it is considered relatively safer and more socially organized. Ferguson's housing stock is affordable enough to be attainable for those wanting to escape the city and enjoy a more suburban lifestyle. However, as suburban growth in the St. Louis region moves westward, so do those who can afford to move westward and build new homes. This leaves inner ring suburbs such as Ferguson to those who are in search of affordable housing options outside the city, but also bringing "physical and social issues" associated with city life with them. These are issues often unwanted to those who presently reside in the suburbs because they challenge the current social norms. McGraw (2015) argues, "many Whites won't publicly say, but is in the forefront of many minds (especially older ones), is that suburbs like Ferguson came to prominence in the late 1940s and early 1950s as a result of White flight. The thought process follows that we (Whites) came to places like Ferguson to get away from them (Blacks), and now they've followed us. So we (Whites) had to move further out." I would extend the argument, based on the results of this research study that this also applies to older adult Black residents within a certain class. They moved from the city, and they don't want people with what they view as incompatible views following them and damaging the social norms of the neighborhood. There is not only a racial division in Ferguson, but a class division as well.

In an interview conducted by Goering (1994) from the Chicago Tribune, Louis Masotti, a former Northwestern University specialist on the suburbs suggests that communities undergo various stages, just as individuals do throughout their life. As cities begin to experience difficulties and infiltrate neighboring regions, it is the innermost suburbs who will first feel the effects. Problems of the city include the negative connotations associated with poverty such as "crime, violence, drug use, homelessness, abandoned properties, and a lack of commercial institutions" which further neighborhood collapse (Kovarik 2008). Although one may think of segregation in terms of geographic racial issues (Whites live there, Blacks live here), segregation can take on many other forms. According to Cooperman (2014), "geographic, cultural, and economic" segregation have all become the norm in the St. Louis area. "North of Delmar Boulevard, St. Louis neighborhoods are about 95 % Black. South of Delmar, they are almost two-thirds White (and the median household income is $25,000 higher) (Cooperman 2014)." Long after racial discrimination was deemed unconstitutional, St. Louis has "chopped itself into bits, remaining socially and economically segregated (Cooperman 2014)."

However, research participants also felt that the City of Ferguson was taking some very proactive steps to control the cycle of renters in their community and encouraging them to become better renters.

> The city is…taking some very proactive steps in rental housing, to control the cycle of people moving in as renters that maybe haven't had a background in home ownership or maybe their extended family is all the culture that they know. Sometimes that brings in people…that bring bad things in with them… That was very proactive and very wise of our city leaders.

In 2006, Ferguson launched a program to hold landlords accountable. This occurred because of the apprehension local residents expressed about the large amounts of rental property appearing in their neighborhoods. Landlords must properly maintain their properties. Neglected properties must be inspected by an ASHI (American Society of Inspectors) inspector at the landlord's expense to make sure code violations are corrected (Missouri Municipal League 2006). Unfortunately, older adults also acknowledged that some rental homes were being walked away from because the expense of bringing the homes up to code was costlier than what the home was worth. Thus, it leaves a further blight on the community as a vacant and abandoned home.

Rephann (2009) outlines four thoughts that align with the Social Disorganization Theory and acknowledges that various arguments exist as to why homeowners are less prone to generate crime and become a victim of crime. First, homeowners do not move because of the associated costs (Dietz and Haurin 2003; Rohe and Stewart 1996). Also because of their longevity in the neighborhood, they have established social connections (Rohe et al. 2002; DiPasquale and Glaeser 1999; Rohe and Stewart 1996). Second, because they invested in the community, they also invest in property maintenance and security for their home. This also intensified "their sensitivity to declining property values and 'quality of life factors' that weaken these values (Dietz and Haurin 2003; Rohe and Stewart 1996)" such as lack of maintenance and abandoned homes. "Third, homeownership has been connected to better child outcomes (Dietz and Haurin 2003; Rohe et al. 2002; Harkness and Newman 2002)." Youth are less inclined to commit crimes because they are living in a stable environment in which they are not transiently moving frequently. Fourth, homeowners have better health, mentally and physically (Dietz and Haurin 2003; Rohe et al. 2002), responding more robustly to stress, and are "less likely to react violently or unpredictably." In this research study, all participants were homeowners. This was likely an enormous factor for older adults in how they viewed renters and rental property in their neighborhoods.

Physical Disorder

In addition to the presence of social disorder in the community, physical disorder was also demonstrated by research participants through dilapidated and overgrown vacant properties (homes, businesses, and land), negligent maintenance, and an increased problem with litter and illegal dumping. Research participants viewed activities such as littering as distasteful and disrespectful to current and future generations. Physical disorder such as lax maintenance, littering, and illegal dumping were also described as contagious because their mere presence warranted unspoken permission to other individuals who pondered trash dumping. While social disorder can be explained through the Social Disorganization Theory, physical disorder can be described in part through the fundamentals of the Broken Window Theory and the Defensible Space Theory. The Broken Window Theory describes that, "if a window in a building is broken and is left unrepaired, all the rest of the windows will soon be broken…one unrepaired broken window is a signal that no one cares…" (Kelling and Wilson 1982). This leads to serious social breakdowns in the community and can accelerate physical and social deterioration in the community. "A piece of property is abandoned, weeds grow up, and a window is smashed. Adults stop scolding rowdy children; the children, emboldened, become more rowdy. Families move out and unattached adults move in" (Kelling and Wilson 1982). This theory also applies to negligent maintenance in the community, littering and illegal dumping, and the further destruction of already vacant properties in the community. These are properties that may have become vacant for some other reason, such as foreclosure or abandonment, but when others learn that they are empty they begin to become repeated targets for theft, vandalism, and other illegal activities.

Maintenance

Coinciding with the concepts of the Broken Window Theory, several older adults felt that residential maintenance had a contagious nature to it. Everyone in the neighborhood was responsible for upholding the social norms of the neighborhood. This meant following city maintenance codes and setting an example for those nearby. Those who violated the norms were outspokenly providing permission to those who had similar intentions, and setting the standard for a snowball effect of potential lax maintenance principles in the community.

While research participants may have expected local police to enforce both formal law and informal social community standards, they expected the City of Ferguson and their code enforcers to ensure physical order and hold individuals responsible for the condition of their properties. They were unpleased when they felt that maintenance codes were not uniformly enforced that enforcers were being careless with their responsibilities, and that homeowners did not take fines associated with code violations seriously. Research participants reported code violations to the City of Ferguson when they felt that those violations were particularly offensive and had not been corrected in a reasonable amount of time. Other maintenance issues did not necessarily violate any city maintenance codes, but seemed to violate either the social norms of the neighborhood or the personal standards of the person who found the issue offensive. Many issues become quite subjective, such as a concern about one residents distasteful "bucket garden." Therefore, what may appear offensive to one person may not be at all offensive to another; especially that person who is violating the perceived offense.

Littering

Nearly all older adults who participated in the research study were concerned with the presence of littering and illegal dumping in their community. Problems with illegal dumping relate to the Broken Window Theory in which older adults felt that if littering and dumping were not controlled, others who had similar intentions would also use properties for dumping. This is because it appeared that no one was taking ownership of the property and did not care about the condition of the space.

> The hardest thing is to keep it cleaned up. If you dump something, get it cleaned up right away so that other people don't think, 'oh, we can bring our stuff and dump it there.'

Several older adults expressed that they had taken it upon themselves to improve the image of their community and reinforce positive social norms by picking up trash. They thought that if people became familiar with clean areas they would be less likely to litter because it appeared that someone cared about the area. Unfortunately, several older adults also admitted significant frustration because the process of cleaning up trash seemed like an endless task in their community.

> ...My husband and I are big about littering. He is going to be known eventually as the 'trash man cometh'...We walk daily...We have different routes...As we walk he picks up trash...It is totally amazing what people will throw out of the cars or leave in the parking lots...

Natural surveillance and territorial reinforcement described as part of the Defensible Space Theory also help define the issue with illegal dumping and littering. As related to illegal dumping and littering, natural surveillance and territorial reinforcement seemed to go hand in hand in—when one was present, so was the other. At the same time, when one was lacking, it also impeded the other. Natural surveillance is defined by Newman (1972) as "the capacity of physical design to provide surveillance opportunities for residents and their agents (p. 78)." On the other hand, Newman (1972) describes territorial reinforcement as "the capacity of the physical environment to create perceived zones of territorial influences (p. 51)."

Particularly, as related to illegal dumping, disposal of larger items occurred in areas where opportunities for natural surveillance were deficient. People who illegal dumped items went to places where they were unconcerned that people would see and report their actions; areas hidden by brush and overgrowth, areas with poor lighting, and areas where the rear of homes back up to the street instead of facing the front of the street.

> Just paper…trash…This is the side that people dump the most…because the residents can't see…If they would clear it out that would stop…They would think people are watching… They wouldn't be dumping stuff…

Illegal dumping also appeared to occur in areas that lacked territorial reinforcement. These areas included creek beds screened off by brush, absent fences or warning signs, and vacant land in which no ownership of the space was exhibited. This lack of ownership over the land breeds a condition in which those who are destined to abuse the area, will do so.

> …by Maline Creek. It appears to have already been used for a drinking and dumping area…beer cans…a tire here by the culvert…Whoever is responsible for the creek access by the road has not taken precautions…to minimize…trash dumping…Maybe a fence should have gone up…Somebody's taking advantage of this area. So how do we minimize that from occurring?

Finally, another older adult spoke about regularly cleaning up the front of her business and defining her personal territory.

> …a lot of dumping has stopped…People see a clean area and they are less likely to litter… When it's busy traffic, that's when I clean my windows. It's when I clean my doors. I'm out there sweeping…I want people to know…

Not only was she setting the social norm for littering on her property, she was also letting people know that someone owns, cares for, and is watching out for their own personal property. She demonstrated a subtler approach to conveying the message that someone is looking out for the well-being of his or her property and the neighborhood.

Vacant Property

Older adults were exceedingly concerned with vacant property in their community. This included vacant homes, businesses, and lots. Often, vacant homes included plywood boards over doors and windows to keep vagrants either from breaking all of the windows and doors and entering the house, or to cover up damage that had already occurred. Research participants felt that the presence of boards on the doors and/or windows of a house looked much worse than a vacant house without boards; possibly attracting the type of people that they were trying to keep out. One research participant demonstrated with pictures that if the necessary plywood was painted to match the house, it made it less evident that the house was empty. Also, older adults agreed that in many instances it might be difficult to even determine if a home is sitting empty except for key obvious signs. The presence of plywood over the doors and windows, untrimmed and overgrown landscape, and code violation stickers on the windows and doors became obvious signs of an empty home. Despite this, it seemed that older adults were less concerned with the presence of crime associated with these properties and more concerned with the blight they created in the neighborhood, how property values were affected, and whether new homeowners would be attracted to the neighborhood. Older adults felt that there really was not much to steal in a vacant home, but acknowledged that copper theft had been a problem. Older adults were also concerned that no one was being held accountable for abandoned properties and that the City of Ferguson had not taken some sort of action on vacant properties. However, the City of Ferguson took some initiative against problem properties and began to tear down some of the vacant homes and businesses during the course of the research study.

Several older adult males stated that they had taken initiative to care for vacant homes near them. This helped to prevent their personal property from looking unpleasant and to reduce the effects of possible crime and vandalism to the vacant property. This in turn increased territorial reinforcement and natural surveillance because: (1) it appears that someone is taking ownership over the property, and (2) someone is looking out for the well-being of the property and reporting suspicious activity. Thus, reducing the vulnerability the home may experience to crime and vandalism.

Spelman (1993) suggests that "the prevalence of abandoned property has been found to be associated with greater crime" (cited by Rephann 2009). According to Kraut (2012), vacant properties offer places for criminals to hide and generate crime, symbolize a declining neighborhood in which there is a lack of control and criminal activity is tolerated (cited by Branas et al. 2012). Becoming an indication that no one is concerned about their neighborhood, increased amounts of empty structures decrease the value of properties and advances further vacancy. The visual image of vacant properties also damages the neighborhood far beyond any calculated dollar amount (National Vacant Properties Campaign 2005). The United States Department of Housing and Urban Development (2014) report that "vacant and abandoned properties have negative spillover effects that impact neighboring properties and, when concentrated, entire communities and even cities. Research links foreclosed, vacant, and abandoned properties with reduced property values, increased crime, increased risk to public health and welfare, and increased costs for municipal governments."

Critical Issues

The City of Ferguson (2014) is working to improve upon some of these issues as outlined in their annual budget that "acts as the city's most important financial tool to establish goals for the upcoming fiscal year." According to the 2014–2015 annual budget, "in 2012 the Ferguson City Council established a list of critical issues that should help drive the budget preparation process each year, to ensure we utilize our limited resources to their fullest extent by focusing on strategic short and long-range goals." The current list of critical issues includes:

- ***Housing***—*to promote home ownership, assist landlords in attracting quality renters, improve property maintenance and better track and manage vacant properties.*
- ***Employee Relations***—*to improve on-the-job satisfaction of City employees through better expressing appreciation and improving communications so as to allow for a better understanding of community goals.*
- ***Attracting and Retaining Residents and Businesses***—*to improve upon services provided to our existing residents and business owners and to develop better strategies for the promotion of our available residential and commercial real estate.*
- ***Infrastructure***—*to continue to focus on improving city streets and sidewalks as two of our primary capital improvement expenditures that aid in fully connecting our community.*

Three of the above critical issues (housing, attracting and retaining residents and businesses, and infrastructure) outlined in the budget parallel some of the concerns that older adults expressed through their photographs and interviews. It appears that the City of Ferguson is taking an active interest to enhance the quality of life for residents and remedy some of the low levels of social disorder that lead to the perception that their community may not be a choice place to live.

Place Attachment

According to Lowe and Altman (1992), place attachment, as defined as one's emotional or affective ties to a place, is generally thought to be the result of a long-term connection with a place (cited by Smaldone 2006). As Smaldone (2006) explains, "this is different from saying a place is special because it is beautiful...one can have an emotional response to a beautiful (or ugly) landscape or place, but this response may sometimes be shallow and fleeting." Manzo and Perkins (2006) acknowledge that place attachment can also be disrupted by such things as "crime, relocation, and environmental disasters (Brown and Perkins 1992)" and "cause feelings of loss and alienation (Hummon 1992)." Manzo and Perkins (2006) also suggest that, "tapping into such feelings and reactions to disruption can, if properly recognized and understood, help mobilize citizen participation to rebuild a community. Conversely, if such feelings and experiences are not well addressed, disruptions could divide a community."

Although older adults were concerned with the presence of low-level symbols of physical and social disorder in their community, they identified many positive aspects about their community that helped to counter the effects. Many older adults had been lifetime or longtime residents of the community. They had an attachment to not only their own individual home and neighborhood, but were invested in the betterment of their entire community. They were proud of their many beautiful parks, farmer's market and community gardens, historic homes, and new blossoming businesses and investors. It appears that place attachment in conjunction with older adults physical and social engagement in the community counter the effects they could experience from the presence of social and physical disorder in the community.

Environmental Docility Theory

Older adult participants demonstrated that they are sufficiently capable of dealing with the effects of low levels of environmental disorder in their community. For these older adults, the personal fear of victimization does not appear to be present. The fear that is present is one in which they are afraid that these low levels of physical and social disorder are compromising social norms and thus symbolizing the future deterioration of their community. Their attachment to place combined with their present ability to be physically and socially engaged in their community has made them sufficiently capable of dealing with the environmental effects of crime and fear of crime in their community. If their present physical and social engagement in the community were to cease, it is possible that they may not be able to sufficiently deal with increased environmental press from fear of crime in their community.

However, in analyzing the statement of one research participant, it is appropriate to connect the concepts regarding the Environmental Docility Theory with the personal mechanisms that older adults utilize to compensate for decreased competency to control particular environmental pressers that may be affecting them.

> I need to say this about the audience that you have been seeking out. When your years get less, you've got more behind you than you've got in front of you. We all start feeling our

> mortality a little bit and that's frustrating...When we were younger we thought we were more tolerant...now we are finding out that we're not. We like to have more control over something. So a lot of the belly aching you hear it seems related to the fact that we have less and less control over our lives. Well, the perception is that control is slipping away from us. And this is belly aching about something like this is a good relief.

As indicated through this research participant's statement, older adults sense the decline in their competency level and as a result they begin to compensate by further controlling that which they still have some power. As competency diminishes, they work to control the effects of specific environmental pressers that are appearing to become unmanageable and uncontrollable. Physical and social norms and standards that are changing beyond what these older adults currently feel is within their scope of acceptance. Physical and social standards which may very well become conventional practice to a more demographically diverse community. As this research participant indicted, older adults attempt to preserve control over the perceived physical and social community values by "belly aching." This research study became an appropriate outlet to voice concerns about the deterioration of physical and social characteristics in the community. That which they can still control will not become an environmental press too monumental to overtake their personal competency. A personal competency level that is potentially diminished, as the environment spirals out of control, becomes too unbearable for one to physical, socially, and emotionally manage.

This statement may have also unintentionally served as a premonition for the future occurrences that took place in the community in August, 2014. The social norms of Ferguson have begun to change as the demographics of Ferguson have changed. Those in power potentially work to grasp even more tightly onto the rules and regulations which would keep the powerless (those deemed unmanageable and uncontrollable) from spinning the current social norms out of control. This creates a struggle for power in the community; feeding distrust, disrespect, and further social disorganization.

Comparison of Demographic Data

Demographic data for the 15 research participants was collected to understand if any specific characteristics were associated with a higher perception of crime. In this research study, gender did not seem to have a specific impact on whether older adults were more fearful of crime in their community. However, it did appear that females took a more active interest in reporting specific incivilities such as maintenance violations to the City of Ferguson. In contrast, males reported taking a more active interest in becoming directly involved in the upkeep of specific properties. For example, all of the male research participants communicated that they had cared for vacant properties next door to or near their own home. This included mowing grass as well as keeping a watchful eye to make sure the home had not been damaged by criminals or through weather-related events.

The age groups of older adults can be categorized into three major subpopulations. Those who are age 65–74 are considered the "young-old," 75–84 are "old," and those who are age 85+ are the "oldest-old" (Transgenerational Design Matters 2009). In this research study, five research participants were between the age of 60 and 64 years of age. Therefore, they technically do not even meet the definition of someone who would be considered the youngest of the old population. Nine research participants were between the ages of 65–74, thus meeting the criteria for the youngest of the old population. Two research participants were between the ages of 75–79, meeting the criteria for the "old" population. None of the older adults who participated in this research study could be included in the "oldest-old" age category. Therefore, it is possible that this could also have an effect on their perceptions of crime in their community. Those who are younger are also more likely to be physically and socially active, assisting in combating the effects of crime perception in their community.

The race of the research participants did not seem to affect whether or not they were more fearful of crime in their community. However, as previously noted, some research participants, depending on their race seemed to view the quality of particular businesses and the type of people they attract differently. Additionally, it was only older White females who identified that they felt that minorities were responsible for increased community crime, littering, as well as other issues symbolizing the decline of their community's social values and norms.

Although specific issues might relate to particular neighborhoods throughout Ferguson, it was not discovered through the interviews with older adults that one neighborhood was more problematic than another. However, two older adults acknowledged that there had been various problems at the apartment complexes off West Florissant Avenue in Ferguson. In addition to their own neighborhood, older adults also ventured outside their individual neighborhood to the main business districts of Ferguson as well as other nearby neighborhoods and parks. However, it appeared that older adults mainly took photos of the older part of Ferguson along South Florissant and did not take any photos of the businesses along West Florissant Avenue.

Whether or not one lives alone is a determining factor in crime perception, especially for female research participants. However, of the four research participants who were males, only one lived alone. One female research participant suggested that she felt that age had little to do with whether or not one was more likely to become a victim of crime. She felt that living alone was a more significant indicator of victimization. This was because it is easier for potential criminals to be able to tell when no one is home, that one individual could be more easily taken advantage of, when others were living in the home they were looking out for one's well-being, and paying attention to when they were supposed to be home.

Only one research participant indicated that she considered herself a person with a disability. However, she did not indicate on the Likert type scale to what degree she felt that her disability made her more susceptible to crime. However, based on the interview, she appeared to be a rather active individual so the potential disability did not appear to inhibit her current activity level. Because all of the older adults indicated that they did not consider themselves a person with a disability, it is possible that this also contributed to their low fear of crime in their community. A disability, particularly a physical disability that prohibits one from being physically mobile, would impede their physical and social activities, thus potentially resulting in an increased fear of crime in their community.

Connection of Research to Michael Brown Shooting

Data collection for this research study concluded in early June 2014, nearly 2 months before the fatal shooting of Michael Brown by Police Officer Darren Wilson in Ferguson, Missouri. Embarking upon the research journey in Ferguson, one never could have imagined that the unfortunate events of August 2014 that lead to the death of Michael Brown would have taken place. These events sparked outrage in the community and throughout the nation; leading to protests, riots, violence, and destruction. Ferguson has become part of history and will continue to shape the way everyone in the nation thinks about race relations in their own communities, shaping new laws, and public policy in the United States.

Two key pieces identified in this research study also relate to the death of Michael Brown. One issue dealt with the frustration that older adults experienced when teens walked down the middle of the street instead of utilizing the sidewalk in their neighborhoods. This was an issue identified by both Black and White female research participants. One Black female also expressed her frustration with teens playing basketball in the street instead of utilizing the local park nearby. This was a complaint often expressed by residents in other various neighborhood watch groups as well. Although walking down the middle of the street or playing basketball in the street may not be a serious criminal offense, it is seen as one that violates the social norms identified by the residents of the neighborhood. Even during neighborhood watch meetings, Black residents of the Nesbit Newton neighborhood (which sits adjacent to the Canfield Green Apartment Complex) were encouraging law enforcement to take action against loitering teens in their neighborhood. It seems sometimes that a police officer's job is not just to "enforce the law," but to be a general peacekeeper in the community in respect to the social norms and values established by the residents themselves. Officer Darren Wilson's initial contact with Michael Brown concerned a confrontation in which the teen was asked to refrain from walking down the middle of the street. However, as it appears in this case, some of the social norms and values of the community have possibly begun to shift, particularly in and around the Canfield Green Apartment Complex.

Several older adults also acknowledged that the apartment area where Michael Brown was shot had been problematic in the past. However, they also felt that the City of Ferguson had taken some very proactive measures to assist in keeping those apartment areas safer. Except for the acknowledgement of the apartment complexes off West Florissant Avenue, it is important to note that most of what older adults in this research study spoke about related to the businesses and neighborhoods along the more central arteries of Ferguson and South Florissant Avenue. Research participants spoke very little about and took no photos of the areas along West Florissant Avenue. Conversely, much of the media has based its view of the entire City of Ferguson on what has happened at the apartment complexes off West Florissant Avenue.

Attorney General Report on Ferguson Police Department

This research study did not specifically examine the law enforcement practices of the City of Ferguson. The main interaction with the Ferguson Police Department was during neighborhood watch meetings, in which they were supportive of the research study, reassuring older adult's that the practices set forth within the autophotography portion were within the rights of the law and that they felt the risk of repercussions were rather low from taking pictures of physical elements and places that represented crime in the community. Interaction with other Ferguson city employees was also positive. Mayor James Knowles expressed interest and support, asking for further information so that he could adequately explain the research study to citizens who had articulated concerns. The city manager John Shaw also asked that the research study be presented at a Ferguson City Council meeting so that members of the public could ask questions directly about the research. The research was presented, but at that time, no questions were asked from the general public.

However, it is important to connect some of the findings from this research report to remarks from the United States Attorney General Report (United States Department of Justice Civil Rights Division [USDJCRD] 2015) on the investigation of the Ferguson Police Department in March, 2015. According to the report, "city officials have consistently set maximizing revenue as the priority for Ferguson's law enforcement activity" (p. 9). Specifically, Judge Brockmeyer approved a fine increase for those who have repeat housing violations (p. 14). The report does not seem to provide specifics on such housing violations. However, as verified through the results of this research study, older adults in Ferguson had some serious concerns about the condition of certain properties and absentee landlords who left their properties in disrepair. Research participants supported Ferguson's efforts to tackle neglected and abandoned properties. However, they did acknowledge that additional fines and fees seemed to have little effect on the condition of the properties because fines were not taken seriously by homeowners.

The Attorney General report also draws attention to the Ferguson Police Department's use of "reasonable-suspicion-based pedestrian stops" (p. 18). According to the report, the Ferguson Police Department had been abusing such policy by detaining individuals wandering about without "articulable reasonable suspicion" (p. 18). As previously indicated, research participants, both Black and White, communicated their concern with groups of teens and young adults wandering the streets. In a neighborhood watch meeting, community members of the predominately Black Nesbit Newton neighborhood specifically asked police officers to questions groups of congregating teens in their neighborhood. The Ferguson police officer present at the meeting explained to residents that it was unlawful to question groups of people without reasonable suspicion because such a practice was susceptible to profiling (racial, gender, etc.) As explained previously in this research study, there is often a fine line between enforcing the law and recognizing the social norms and values established by the community. Although it may not be illegal to congregate in groups, community members had identified the action as suspicious and disruptive; specifically asking that the police force to take action. In an example given in the Attorney General's report, a Ferguson police officer arrested a man for "Manner of Walking in Roadway" because he was using profanities and dancing in the street (p. 25). Again, two female research participants, one Black and one White indicated their frustration with young adults walking down the middle of the street rather than using the sidewalk "screaming at one another or fighting" and viewing such actions as "intimidating to seniors."

Community Development Strategies

Some key concepts of community development include building strong and trusting relationships with partners, enhancing the capacity of community members to make critical decisions, designing initiatives that are all inclusive, and developing strategies to help organize individuals to solve community problems. This will aid community members to develop a sense of purpose, enhance their capacity to be vital and active citizens, and become responsible for the community sustainability. Littrell and Littrell (2006) explain that "the profession of community development is not interested in the establishment of clients….Community development is interested in forming equal partnerships over time that are based on the assumption that people are capable of becoming increasingly able to give purposeful direction to the future of the communities in which they live and function" (p. 7).

Partnerships

A critical aspect of enhancing capacity in the community involves implementing a democratic and inclusive decision-making process that will help increase the social wellbeing of everyone in the neighborhood. Wheatley (2006) explains that the "world of relationships is rich and complex" (p. 35). Wheatley (2006) references Bateson (1980) in explaining that we should "stop teaching facts" or "the 'things' of knowledge—and focus, instead, on relationships as the basis for all definitions" (p. 35). Relationships provide interdependent connections in which everyone involved develops the capacity to impact one another through community development research (p. 35).

People must be directly involved in creating and developing a community change strategy. This includes involving a critical mass of people, or sufficient numbers of people so the idea will be adopted by the whole population (Axelrod 2002, p. 52). Recruiting a critical mass of people for potential partnerships can be accomplished in various ways and help the community take ownership of issues and care about the outcome and sustainability of initiatives (p. 53). When forming partnerships, it is also important to widen the circle of involvement to enlarge discussions about the initiative, provide opportunities for co-created plans, and generate the prospect for sharing new information and ideas. Partnerships should include those who have new information, those who will be impacted by the initiative, those who might oppose the initiative, those who might be responsible for the outcome of the initiative, and those who have the authority to assist in implementing change.

Capacity

Mattessich et al. (1997), define social capacity as "the extent to which members of a community can work together effectively…develop and sustain strong relationships, solve problems and make group decisions, and collaborate effectively to identify goals and get work done" (p. 61). According to Mayer, Rainbow Research (1994), a community can fortify its capacity by uniting the "influence of a community's commitment, resources, and skills that can be deployed to build on community strengths and address community problems" (cited by Mattessich et al. 1997, p. 62). By building an individual's social capacity they will in turn become more effective leaders in their community. Empowering individuals to think about the assets that they provide to the community will enable them to be more effective citizens and encourage them to take action. The community will be more powerful because of the contribution being made. Strong communities are places where the capacities of local residents are identified, valued, and used (Kretzmann and McKnight 1993, pp. 13–14).

Inclusivity

According to Littrell and Littrell (2006), broad-based inclusion refers to "including people and groups with many different backgrounds, ideas, and assets. Inclusion is achieved by seeking out the diverse opinions of people whether they belong to seemingly homogeneous groups or not" (p. 56). Issues identified and resolved should include a broad representation of people and their ideas. This also means that those involved in any community development project should try to understand different viewpoints, discover new partners in the community, and engage those with diverse ideas and backgrounds. "Truly meaningful participation requires that all concerned and affected stakeholders are provided the information and resources they require to influence and contribute to the decision-making process…The issues of who participates, when they participate and how they participate are critical to achieving fairness, efficiency, and stability in decision making" (Roseland 2005, p. 191). Broad representation should extend past descriptors such as gender, age, class, race, and ethnicity. Diverse viewpoints are more important than diversity itself and deliberation helps people to make decisions and solve problems (Matthews 1999, p. 220). The challenge may be discovering, including, and dealing with diverse viewpoints and values that all individuals in the community have (Littrell and Littrell 2006, pp. 71–72). One would hope that by including a cross section of residents to participate in the deliberative process involving the community, residents would be able to engage in a meaningful initiative that includes diverse viewpoints and relieves some of the stereotypes and fears. While it might be challenging to deliberate with those who are culturally different, diversity of experiences will be enriched and value added to support an initiative that truly feels community-wide. A diversified group of stakeholders will help create community cohesiveness and pride that may be more powerful in combating fear of crime than fixing crime itself.

Community Policing

A suggested community development strategy, as outlined in the Attorney General's Report (USDJCRD 2015), is known as community policing. Community policing affords law enforcement and community individuals the prospect of working in tandem to determine the sources of crime throughout Ferguson to work together to discover what causes crime and disorder throughout Ferguson and then prioritize the most effective options for police to address community crime (p. 86). "The focus of these strategies…is on crime prevention rather than on making arrests (USDJCRD 2015, p. 86)."

According to the United States Department of Justice—Bureau of Justice Assistance (1994), there are two main components to community policing. The first of the components is community partnership. This involves, "establishing and maintaining mutual trust" with a "broadened outlook" that "recognizes the value of activities that contribute to the orderliness and well-being of a neighborhood" (p. 15). The second component to community policing is problem solving. Problem solving involves much more than merely preventing and abolishing crime (p. 17). It is "based on the assumption that "crime and disorder can be reduced in small geographic areas by carefully studying the characteristics of problems in an area, and then applying the appropriate resources…" and on the assumption that 'individuals make choices based on the opportunities presented by the immediate physical and social characteristics of an area. By manipulating these factors, people will be less inclined to act in an offensive manner'" (p. 17–18). Community policing has the potential to generate positive encounters between police officers and residents of the community because they have created a partnership in which the capacities of all affected and interested individuals have been considered in the public safety decision-making process. People on the community feel safer when they have assisted in assuming some of the responsibility as part of the community policing process (p. 87).

This research study did not work a great deal in partnership with the local police to identify and address crime and disorder throughout Ferguson. However, this study did work with a segment of the population in Ferguson who were able to both photograph and verbally express some of their concerns in regard to some of the physical and social contributors to crime in their community. As part of the research study, many participants expressed that they hoped the results would be turned over to the City of Ferguson, so that some of the problem areas could be addressed to make Ferguson a safer and more desirable place for everyone to live.

Conclusion

Communities are ever evolving, in which the residents of those communities mold and shape its history. Jane Jacobs (1961) describes cities as "a problem in organized complexity" (p. 422). Cities present "situations in which a half-dozen or even several dozen quantities are all varying simultaneously and in subtle interconnected ways. Cities do not exhibit one problem in organized complexity, which if understood explains all. They can be analyzed into many such problems or segments, which are also related with one another. The variables are many, but they are not helter-skelter, they are interrelated into an organic whole" (p. 433). As Jacobs explains, problems presented in "real life" are not simple problems (p. 434).

The residents of Ferguson, or at least the ones I interacted with as part of this research study, are very dedicated individuals who will continue to make their community a better place for themselves and future generations. They too, will overcome this, learning from the challenges, and utilizing these events to strengthen the fabric of their community. Ferguson is where people have grown up, raised their children, and entertained their grandchildren. Ferguson has fantastic restaurants, shops, and other activities to engage in. The older adults in Ferguson demonstrated what being part of a community really meant. Being part of a community means loving it despite its quirks and peculiarities. It is dealing with the challenges as they come along and working together to find common solutions to make the community better for all who live, play, and work in Ferguson. Those who inhabit Ferguson want to become part of the solution, rather than part of the problem. They will work together as a community to help one another and establish peace as Ferguson begins the healing process.

Considerations for Future Research

The data for this research study was collected prior to the events of August 2014 that resulted in the death of Michael Brown as well as the protests and riots that occurred in the following months. Therefore, it would be quite appropriate to conduct a follow-up research study to understand if older adults' perceptions of crime in Ferguson, Missouri have been altered as a result of the events of August 2014.

This research study had some limitations as every research study does. One included the relatively small sample size. This is common with qualitative research studies, which provided a depth of knowledge and understanding unmatched by a quantitative research study. However, if the interest were to gather a more representative sample and improve the validity of the research results that could then be applied to a broader population, it would be appropriate to conduct a research study utilizing quantitative research methods. Combining the two research method types to develop a mixed-methods approach would also be appropriate for this type of research study. This would combine the particulars of specific research participant accounts formed through qualitative research methods with the generalizability of a larger sample size found in quantitative research studies. Given the events that occurred in August 2014, qualitative research proved to be an appropriate method of data collection. Older adults provided insightful observations about issues present in their community that would have remained undetected in the results of a quantitative research study.

References

Austin, D. M., & Sanders, C. (2007). Graffiti and perceptions of safety: A pilot study using photographs and survey data. *Journal of Criminal Justice and Popular Culture, 14*(4).

Axelrod, R. H. (2002). *Terms of engagement: Changing the way we change organizations*. San Francisco, CA: Berrett-Koehler Publishers Inc.

Bell, K. (2013, October). Pricey hair extensions are hot trend in St. Louis, attracting smash-and-grab thieves. *St. LouisPost-Dispatch.* Retrieved from http://www.stltoday.com/news/local/crime-and-courts/pricey-hair-extensions-are-hot-trend-in-st-louis-attracting/article_9c581f13-a2e1-53b8-8f9c-e8ffb6fbc28e.html

Bogan, J. (2014, October). As low-income housing boomed, Ferguson pushed back. *St. Louis Post-Dispatch.*

Branas, C. C., Rubin, D., & Guo, W. (2012). Vacant properties and violence in neighborhoods. *International Scholarly Research Network Public Health,* 2012, Article ID 246142.

City of Ferguson. (2014). *Annual operating budget: Fiscal year 2014–2015.* Retrieved from http://www.fergusoncity.com/DocumentCenter/View/1701

Clare, J., & Plecas, D. (2012). A framework for conducting annual community safety audits: An in-house methodology for police departments (Working paper No. 41). *International Police Executive Symposium and Geneva Centre for the Democratic Control of the Armed Forces.* Retrieved from http://www.dcaf.ch/Publications/IPES-DCAF-Working-Paper-Series

Cooperman, J. (2014, August). St. Louis: A city divided. *Aljazeera America*. Retrieved from http://america.aljazeera.com/articles/2014/8/18/st-louis-segregation.html

Donley, A. M. (2008). *The perception of homeless people: Important factors in determining perceptions of the homeless as dangerous* (Doctoral dissertation). University of Central Florida. Retrieved from http://etd.fcla.edu/CF/CFE0002261/Donley_Amy_M_200808_PhD.pdf

Dionne, E. J., Jr. (Ed.). (1998). *Community works: The revival of the civil society in America*. Washington, D.C.: Brooking Institution Press.

Felonneau, M. L. (2004). Love and loathing of the city: Urbanophilia and urban phobia, topological identity and perceived incivilities. *Journal of Environmental Psychology, 24*, 45–52.

Flournoy, R. (2006, Fall). *Healthy foods, strong communities: Fresh fruits and veggies are good for more than just your health*. National Housing Institute. Shelterforce Online. Retrieved from http://www.nhi.org/online/issues/147/healthyfoods.html

Goering, L. (1994, September). *Inner-ring losing its glow: Close-in towns struggle to survive 50 years after onset of their glory days*. Chicago Tribune. Retrieved from http://articles.chicagotribune.com/1994-0904/news/9409040249_1_suburbs-chicago-places

Jacobs, J. (1961). *The death and life of great American cities*. New York, NY: Random House.

Kelling, G. L., & Wilson, J. Q. (1982, March). Broken windows: The police and neighborhood safety. *Atlantic Monthly*, 29–38.

Kovarik, F. (2008, November). *Mapping the divide: A lifelong St. Louisan grapples with the city's racial disparity*. St. Louis. Retrieved from http://www.stlmag.com/Mapping-the-Divide/

Kretzmann, J. P., & McKnight, J. L. (1993). *Building communities from the inside out: A path toward finding and mobilizing a community's assets*. Skokie, IL: ACTA Publications.

Kubrin, C. E. (2009). Social disorganization theory: Then, now, and in the future. In M. D. Krohn, A. J. Lizotte, & G. P. Hall (Eds.), *Handbook on crime and deviance* (pp. 225–236). New York, NY: Springer Science & Business Media. Retrieved from https://webfiles.uci.edu/ckubrin/SD%20Theory%20Then,%20Now%20and%20in%20the%20Future.pdf?uniq=fn2w4g

Littrell, D. W., & Littrell, D. P. (2006). *Practicing community development*. Columbia, MO: University of Missouri Extension.

Mae, F. (2007). Homelessness in America: Americans' perceptions, attitudes and knowledge. *General Population Survey & City Surveys*. Gallup, Inc. Retrieved from http://shnny.org/uploads/2007_Gallup_Poll.pdf

Manzo, L. C., & Perkins, D. D. (2006). Finding common ground: The importance of place attachment to community participation and planning. *Journal of Planning Literature, 20*(4), 335–350. Retrieved from https://my.vanderbilt.edu/perkins/files/2011/09/ManzoPerkins.2006.Neighborhoods-as-commonground.JPL_.pdf

Mattessich, P. W., Monsey, B., & Roy, C. (1997). *Community building: What makes it work?*. St. Paul, MN: Amherst H. Wilder Foundation.

Matthews, D. (1999). *Politics for People: Finding a responsible public voice* (2nd ed.). Chicago, IL: University of Illinois Press.

McGraw, D. (2015, January). *The complications of our deteriorating inner ring suburbs*. Belt. Retrieved from http://beltmag.com/complications-deteriorating-inner-ring-suburbs/

Missouri Municipal League. (2006). Ferguson responsible landlord initiative. *Innovation Award Catalog of Projects*. Retrieved from http://c.ymcdn.com/sites/www.mocities.com/resource/resmgr/innovation_awards/database.06.10.pdf

National Vacant Properties Campaign. (2005). *Vacant properties: The true costs to communities*. Retrieved from http://www.smartgrowthamerica.org/documents/true-costs.pdf

Newburn, T., & Rock, P. (2004). *Living in fear: Violence and victimisation in the lives of single homeless people*. The London School of Economics and Political Science: Mannheim Centre for Criminology and Crisis. Retrieved from http://www.crisis.org.uk/data/files/publications/LivingInFear_prlim%5B1%5D.pdf

Newman, O. (1972). *Defensible space: Crime prevention through urban design*. New York, NY: Macmillan.

Porter, J. R., Rader, N. E., & Cossman, J. S. (2012). Social disorganization and neighborhood fear: Examining the intersection of individual, community, and county characteristics. *American Journal of Criminal Justice, 37*, 229–245.

Rephann, T. J. (2009). Rental housing and crime: The role of property ownership and management. *The Annals of Regional Science, 43*(2), 435–451.

Roseland, M. (2005). *Toward sustainable communities: Resources for citizens and their governments*. Gabriola Island, Canada: New Society Publishers.

Rothstein, R. (2014). *The making of Ferguson: Public policies at the root of its troubles*. Economic Policy Institute. Retrieved from http://www.epi.org/publication/making-ferguson/

Sampson, R. J., & Groves, W. B. (1989). Community structure and crime: Testing social disorganization theory. *American Journal of Sociology, 94*(4), 774–802.

Sampson, R. J., & Raudenbush, S. W. (2004). Seeing disorder: Neighborhood stigma and the social construction of "broken windows". *Social Psychology Quarterly, 67*(4), 319–342.

Smaldone, D. (2006). The role of time in place attachment. In *Proceedings of the 2006 Northeastern Recreation Research Symposium*. Retrieved from http://www.nrs.fs.fed.us/pubs/gtr/gtr_nrs-p-14/7-smaldone-p-14.pdf

Small, M. L., & McDermott, M. (2006). The presence of organizational resources in poor urban neighborhoods: An analysis of average and contextual effects. *Social Forces, 84*(3).

Transgenerational Design Matters. (2009). *The demographics of aging*. Retrieved from http://transgenerational.org/aging/demographics.htm

United States Department of Housing and Urban Development. (2014, Winter). *Vacant and abandoned properties: Turning liabilities into assets*. Evidence Matters. Retrieved from http://www.huduser.org/portal/periodicals/em/winter14/highlight1.html

United States Department of Justice Office of Justice Programs Bureau of Justice Assistance. (1994). *Understanding Community Policing: A Framework for Action*. Retrieved from http://www.ncjrs.gov/pdffiles/commp.pdf

United States Department of Justice Civil Rights Division. (2015). *Investigation of the Ferguson Police Department*. Retrieved from http://www.justice.gov/sites/default/files/opa/press-releases/attachments/2015/03/04/ferguson_police_department_report.pdf

Wheatley, M. (2006). *Leadership and the new science: Discovering order in a chaotic world*. San Francisco, CA: Berrett-Koehler Publishing Inc.

Index

Note: Page numbers followed by *f* and *t* indicate figures and tables, respectivel

K.L. Fisher-McLean, *Perceptions of Community Crime in Ferguson, MO*,
SpringerBriefs in Criminology, DOI 10.1007/978-3-319-46786-3

MIX
Papier aus verantwortungsvollen Quellen
Paper from responsible sources
FSC® C105338

If you have any concerns about our products,
you can contact us on
ProductSafety@springernature.com

In case Publisher is established outside the EU,
the EU authorized representative is:
Springer Nature Customer Service Center GmbH
Europaplatz 3, 69115 Heidelberg, Germany

Printed by Libri Plureos GmbH
in Hamburg, Germany